NEW SECURITY FRONTIERS

Global Interdisciplinary Studies Series

Series Editor: Sai Felicia Krishna-Hensel
Interdisciplinary Global Studies Research Initiative,
Center for Business and Economic Development,
Auburn Montgomery, USA

The Global Interdisciplinary Studies Series reflects a recognition that globalization is leading to fundamental changes in the world order, creating new imperatives and requiring new ways of understanding the international system. It is increasingly clear that the next century will be characterized by issues that transcend national and cultural boundaries, shaped by competitive forces and features of economic globalization yet to be fully evaluated and understood. Comparative and comprehensive in concept, this series explores the relationship between transnational and regional issues through the lens of widely applicable interdisciplinary methodologies and analytic models. The series consists of innovative monographs and collections of essays representing the best of contemporary research, designed to transcend disciplinary boundaries in seeking to better understand a globalizing world.

Also in the series

Cities and Global Governance
New Sites for International Relations
Edited by Mark Amen, Noah J. Toly, Patricia L. McCarney and Klaus Segbers
ISBN 978-1-4094-0893-2

Order and Disorder in the International System
Edited by Sai Felicia Krishna-Hensel
ISBN 978-1-4094-0505-4

Legacies and Change in Polar Sciences
Edited by Jessica M. Shadian and Monica Tennberg
ISBN 978-0-7546-7399-6

International Order in a Globalizing World
Edited by Yannis A. Stivachtis
ISBN 978-0-7546-4930-4

Global Cooperation
Challenges and Opportunities in the Twenty-First Century
Edited by Sai Felicia Krishna-Hensel
ISBN 978-0-7546-4678-5

New Security Frontiers

Critical Energy and the Resource Challenge

Edited by

SAI FELICIA KRISHNA-HENSEL

Auburn, Montgomery, USA

Routledge

Taylor & Francis Group

LONDON AND NEW YORK

First published 2012 by Ashgate Publishing

2 Park Square, Milton Park, Abingdon, Oxon OX14 4RN
711 Third Avenue, New York, NY 10017, USA

Routledge is an imprint of the Taylor & Francis Group, an informa business

First issued in paperback 2016

British Library Cataloguing in Publication Data
New security frontiers : critical energy and the resource
challenge. -- (Global interdisciplinary studies series)
1. Energy security. 2. Food security. 3. Natural
resources--International cooperation. 4. Geopolitics.
I. Series II. Krishna-Hensel, Sai Felicia.
333.7'17-dc22

Library of Congress Cataloging-in-Publication Data
New security frontiers : critical energy and the resource challenge /
[edited] by Sai Felicia Krishna-Hensel.
p. cm. -- (Global interdisciplinary studies series)
Includes bibliographical references and index.
ISBN 978-1-4094-1979-2 (hbk)
1. Energy security. 2. Energy policy. 3. Power
resources--Political aspects. 4. Energy consumption--Political aspects. I.
Krishna-Hensel, Sai Felicia.
HD9502.A2N496 2011
333.79--dc23

2011036629

ISBN 978-1-4094-1979-2 (hbk)
ISBN 978-1-138-25041-3 (pbk)

Contents

List of Figures *vii*
List of Tables *ix*
List of Contributors *xi*

Introduction 1
Sai Felicia Krishna-Hensel

1 Energy Security and the Revival of Geopolitics: The Russian Case 9
Mikhail A. Molchanov

2 Energy Transitions and International Security in the Twenty-first Century 31
Mert Bilgin

3 A Grounded Comparison of Energy Security in Denmark, Brazil, Bangladesh, and China 67
Benjamin K. Sovacool

4 Global Energy Governance Post-2012: Bridging the Gap Between Energy Security and Climate Policy 93
Meraiah Foley

5 An Economic and National Security Perspective on Critical Resources in the Energy Sector 113
Nayantara D. Hensel

6 Hydropolitics in South American International Relations: A Perspective on Water Governance at the Prata Basin – Between Conflict and Cooperation (1960–1992) 139
Fábio Albergaria de Queiroz

7 A Need for New Institutions? Bilateral Deals over Food Producing Resources and International Co-Operation on Food Security 173
Benjamin Shepherd

Conclusions 199
Daniel S. Papp

Index *215*

List of Figures

4.1	Energy Security and Climate Policy: Bridging the Gap	96
6.1	Distribution of Global Water Resources	139
6.2	International River Basins of South America	140
6.3	World Map from the Perspective of Distribution of Water Resources	141
6.4	Water Availability Versus Population	142
6.5	Classification of RSC from the Patterns of Amity/Enmity	150
6.6	The Prata Basin	153
6.7	Paraguay-Parana Waterway	162

List of Tables

2.1	The Oil Impact in the Twentieth Century	39
2.2	Crude Oil Prices (US Average, $/bbl)	40
2.3	World Marketed Energy Use by Fuel Type, 1990–2035 (quadrillion Btu)	43
2.4	World Marketed Energy Consumption by Country Grouping, 2007–2035	44
2.5	New Seven Sisters, Energy and State	47
2.6	World Energy-related Carbon Dioxide Emissions, (billion metric tons)	50
2.7	Projected Demand for Feedstock Commodities for Bio-fuels in 2020	54
2.8	FAST Matrix: Selected Energy Challenges and International Security	57
5.1	Distribution of Uses for Rare Earth Minerals	117
5.2	Uses of Specific Rare Earth Elements	117
5.3	Uses of Non Rare Earth Materials	127
5.4	Distribution of Reserves of Non-Rare Earth Critical Materials	128
5.5	Percentage Distribution by US Import Sources for Critical Materials	129
6.1	Application of the Theoretical Model of the School of Copenhagen to Hydropolitics	147
6.2	Most Important Hydraulic Projects at the Prata Basin	161

List of Contributors

Sai Felicia Krishna-Hensel, Director, Interdisciplinary Global Studies Initiative, Auburn Montgomery, USA

Mikhail A. Molchanov, Associate Professor, Department of Political Science, St.Thomas University, Canada

Mert Bilgin, Associate Professor of International Relations, Department of Political Science and International Relations, Bahçeşehir University, Istanbul, Turkey

Benjamin K. Sovacool, Assistant Professor, Lee Kuan Yew School of Public Policy, National University of Singapore

Meraiah Foley, Centre for International Security Studies, University of Sydney, Australia

Nayantara D. Hensel, Chief Economist, US Department of the Navy, Washington DC, USA

Fábio Albergaria De Queiroz, Department of International Relations, University of Brasília, Brazil

Benjamin Shepherd, Centre for International Security Studies, University of Sydney, Australia

Daniel S. Papp, President, Kennesaw State University, Georgia, USA

Introduction
Global Security in the Twenty-first Century: The Critical Energy and Resource Challenges

Sai Felicia Krishna-Hensel

The globalizing world is increasingly confronting a new category of security issues related to resource availability and resource depletion. The emerging threat to international security and stability posed by the unreliable availability of essential resources, including water, food, and energy is impelled by increasing demand fueled in part by demographic growth. The competition for energy and food resources is also being driven by the industrialization of the developing world, as well as the demands of the industrialized powers. While industrialized nations continue to consume resources at a disproportionate rate relative to the rest of the world, as populous nations like India and China join in the economic opportunities of a globalizing world, their needs are likely to add to the already crowded field of resource demand and consumption.

The new security arena is best described as a complex matrix, fashioned by interdependencies. This is reflected in the unintentional effect of food insecurity produced through pursuit of diversification in response to the growing energy needs. The development of alternate energy to assuage environmental concerns, as well as in response to depletion of existing resources has impacted food supply and prices as cropland is turned to energy related crops. The resulting increase in food insecurity is problematic to say the least, adding uncertainty to the existing factors of poverty, accessibility, and interruption of supplies through conflict or natural disasters.

The unintentional effects of innovative strategizing represent a new challenge to policy makers. In short, as energy issues combine with other resource challenges, the resource security matrix becomes increasingly intricate.

The modern concern with energy security gained momentum with the "oil weapon" terminology in the aftermath of the 1973 Arab-Israeli war and continued in the period following the collapse of the Soviet Union as oil rich ex-Soviet republics flexed their newly acquired energy power. Resource conflict analysis that accompanied these events focused on the relationship between markets, states, and non-state actors. This line of reasoning continues to characterize the contemporary discussion of energy security. An acknowledgement of the changing security environment and the introduction of new constituents in the causality mix is essential to understanding the present challenges.

In the past 35–40 years, worldwide energy consumption has nearly doubled, driven by population growth, rising living standards, invention of energy-dependent technologies, and consumerism. If these trends continue, global energy consumption is likely to double again by mid-century. The consumption of every major energy source, with the exception of coal, has increased markedly. Without changes in the overall energy mix, fossil fuels will continue to dominate. Fossil fuel dependency, accompanied by a reduction in food production impelled by climate, water resource competition and similar issues will shape a significant proportion of international relations. The rising energy need of the twenty-first century revolves on the availability of oil resources. A real concern is the possibility that global oil production could be interrupted or could reach a ceiling, thereby initiating a rapid increase in the price of oil that, in turn, would trigger a global recession. Economic hardship could lead to conflict over oil supplies. Any reduction of global oil production would have the potential to create havoc both in the oil producing and oil consuming nations of the world. The rapid rise in revenue could also contribute to unrest and civil war in developing countries with oil reserves as the population seeks a more equitable distribution of the country's natural resources. Countries such as Nigeria, Iraq, Iran, etc. for example, all produce oil but present problems ranging from a stable supply to a stable government. Other producers, such as Russia and Venezuela use oil as a lever in international relations to assert power and ensure compliance with their national objectives.

The tension created by the intersection of increasing demand and unreliable supply is likely to manifest itself in a number of confrontational scenarios. The future could see instability and conflict arising from resource scarcity. Increasingly, industrialized states of the world will find themselves in competition with the developing world over critical resources as they seek to sustain their economies and meet the expectations of their populations. Geopolitical issues, new and old, are therefore likely to arise as legitimate stability and supply issues continue to be a major area of concern. Some countries, such as the US, have enormous military expenditure in part to protect global oil areas for their interests. Energy related concerns are impelling a number of other large countries, including China and Russia to be more active in the international arena, prompting fears of a geopolitical cold war centered on energy security. Energy security is one of the most important strategic challenges facing the United States and the world. Energy security cuts across many sectors – economic, environmental and national security. Strategic military thinking in the United States echoed in some measure in Europe has called for greater involvement in energy security issues.

The use of rare earth elements in modern technology has significantly increased in recent years. Many technological devices incorporate rare earth elements. Fiber optical communication, cathode ray tubes, batteries and magnets would be unimaginable without rare earth components. The areas rich in rare earth deposits and production were located in South Africa, India, Brazil, and California. China, however, has come to dominate the world's rare earth supply by producing over 97 percent of the global requirements. This comes at a time of increasing demand and decreasing reliability of supplies. The current concern focuses on the possibility of a shortage as China has indicated a reduction of exports to conserve resources and the environment.

Water resources have historically been drivers in the development of agriculture, trade, urbanization and have contributed to the establishment of powerful states. Competition over shared resources has had the potential for major conflict. With the growth of mega-cities the problem is more acute as demand for water is exceeding the ability to meet the need of millions of urban residents. The danger that water resources will be diverted from agriculture to urban development is one that has special implications for the developing world. These problems are compounded by environmental concerns ranging from quality of the water supply to ecosystem management. Climate changes resulting in drought and surface evaporation contributes to the overall dilemma posed by declining water supplies and burgeoning global demand. Shared resource management is one solution to the equitable distribution dilemma.

The current and projected global population growth estimates are not reassuring for ensuring food security in the developing world. Sub-optimal agricultural productivity based on many factors including increasingly scarce water resources, climate change, technical expertise, and labor, are contributing to uneven food production and distribution. Additional complications involve the absence of satisfactory trade protocols for importing food, as well as long term development strategies. The problem is often most effectively tackled at the community, regional level, where innovative policies and immediate effects are measurable.

This volume brings together a number of these concerns focusing on global resources, as well as, assesses the existing global governance landscape to include current and newly emerging organizations, institutions and policy networks. The chapters address energy, water, and food security and the geopolitical ramifications of resource competition, as well as explore the development of strategic thinking on these issues. These analyses underscore the increasing awareness that this is a critical area of concern in the twenty-first century global security landscape. The essays emphasize both the uniqueness and the magnitude of the new challenges while, simultaneously acknowledging that the current predicament has a contextual history in the power configurations and hierarchy of the international system. The new security environment is both qualitatively and quantitatively distinct, and the resource environment contains both traditional resource categories and a host of new technologically related resources. The globalizing environment creates more unique dilemmas and new centers of political/economic power. Addressing the security environment from a historical, geo-political perspective, and examining specific regional and national case studies, this volume emphasizes the significance of the new security frontiers and explores innovative policy options.

Mikhail A. Molchanov situates the discussion of energy security within a geopolitical context. He observes that geopolitics and energy security represent the paradox between the universal character of energy demand and the nation-state's exclusive right of management of natural resources located within its realm.

A survey of the historical antecedents of the geopolitics of natural resources and the imperialist struggle for oil at the beginning of the twentieth century introduces the discussion. This is followed by an overview of the oil crises of the seventies, and the end of market domination by a limited number of players. The subsequent energy landscape was marked by the institution of an energy regime that was increasingly dependent on the stability of the producing countries. With the advent of globalization, there was an increasing realization that the energy environment had to be viewed from a global perspective on traditional rivalries. The analysis includes a case study that focuses on the contemporary struggle over control of energy resources in Central Asia, Russia and Eurasia. With Russia reasserting control over its energy production and trade, there is a renewed emphasis on its energy policies and foreign economic relations which take on added significance in the debate over energy security in both the regional and global contexts. The chapter concludes with the contention that energy security lies at the intersection of conventional security concerns and environmental imperatives that have emerged from global climate change and global resource depletion. From the perspective of traditional realism, the possession of resources, as well as the ability to utilize them effectively, continues to determine the relative standing and status of states within the international system. The only element that has changed is the evolving definition of strategic resources.

Mert Bilgin explores the relationship between energy transitions and international security arguing that twenty-first century energy security represents a new energy order that is fundamentally different from earlier eras and may be understood from the perspective of energy transitions elaborated as the FAST principles (Feasibility, Accessibility, Sustainability, and Transparency). This approach attempts to differentiate between the actors, challenges, and opportunities that determine the balance between energy transitions and security. The progression of energy resource extraction and use over the centuries as humankind transitioned from primitive energy to fossil fuels and the later development of nuclear energy and bio-fuels provides an insight into the factors underlying the ascendance and decline of states in the international order. Relying on a historical overview, the author concludes that in the absence of a technological breakthrough to satisfy the increasing energy demands, the declining quantities of renewable sources will result in a changing geopolitical landscape. With energy security and economic growth seen as vital components of global or regional power status, global competition will have to be seen in ideological terms as well. The analysis also points to the impact of bio-fuel development on food prices and how that characterizes a forthcoming global challenge.

Benjamin K. Sovacool's chapter suggests that even as we recognize that energy security is important for modern society we are faced with market failure and under-distribution that undermines this recognition. Sovacool argues that energy security is fundamentally reliant on policies and methods to equitably provide inexpensive, dependable, effective, environment friendly, fairly administered and socially acceptable energy services. The chapter, consequently, focuses on identifying policies and programs that incentivize energy security. The author's methodology includes a combination of interviews and field research in seven countries combined with four case studies. The four representative case studies of energy security are electricity supply, transport, household energy use, and off grid energy services. The rationale for selecting electricity supply is based on the assessment that electricity accounts for more than half of global energy consumption each year; transportation is selected since its total energy use is second after electricity. Denmark provides an ideal study for electricity supply because of its high rate of renewable energy diversification, while Brazil is investigated since it, in conjunction with the United States, dominates the global ethanol market. The chapter examines household energy use and off grid energy services in the developing world to contrast energy use in industrialized countries. Noting that compared to electricity and petroleum-based motorized transport in industrialized countries, 3 billion people still rely on traditional biomass fuels for cooking and heating, 1.5 billion have no access to electricity, and 1 additional billion have access to only unreliable electricity networks, Sovacool reviews two innovative programs; solar and biogas units in Bangladesh, and cookstove distribution in China.

Meraiah Foley's chapter is an exercise in applying collective action theory to global energy governance. She suggests that the absence of comprehensive frameworks to manage the global energy market, in particular the global oil market and the security of its supply is particularly unhelpful in light of the importance of energy related issues in the contemporary world. Collective action has long been regarded as a vital element in – and a major barrier to – the successful management of common pool resources. Several studies have used collective action theory to understand the dynamics, and limit the losses, that often occur where multiple companies have claims on a single oil field. It has also been applied to the broader issue of climate change. However, no study has yet applied collective action theory to the regulation of the global oil market. This analysis attempts to remedy this omission by exploring whether collective action theory is a meaningful tool for understanding the lack of global coordination around energy markets, particularly in regards to the demand and supply of oil. It further assesses the relevance of collective action theory in developing a comprehensive framework of energy governance, including the conditions, if any, that must be met for successful collective action in this arena. The premise that energy-producing and energy-consuming states share a common interest in improving efficiencies in the global management of oil markets – pricing; supply and demand; new investments is dependent on an appropriate analytic framework for achieving this goal.

In Chapter 5, Nayantara D. Hensel takes an economic and national security perspective on natural resources. She observes that while global economic interdependence has contributed to a higher level of global prosperity, it has also created dependencies on critical raw materials that are very often imported from abroad. This has national security implications in that an interruption in the flow of these critical raw materials can seriously dislocate strategically vital sectors of the economy of the dependent state. In her study, Nayantara Hensel examines the challenges confronting the United States as it seeks to reduce its dependency on the importation of oil from overseas suppliers, especially those in the Middle East. Specifically, she examines the US efforts to diversify the sources from which it draws critical energy resources, as well as efforts to develop alternative energy efficient technologies, including biofuels and innovative technologies that rely upon rare earth minerals and other critical raw materials. She suggests, however, that, in its efforts to acquire energy independence by developing alternative energy resources to meet the needs of the US economy in the twenty-first century, the US should take care to avoid exchanging one pattern of critical resource foreign dependency for another pattern that also relies on foreign sources for equally critical, though different raw materials. Hence, she argues that the US needs to both aggressively develop domestically based critical resource mining operations, as well as diversify the overseas suppliers of those resources that are key to sustaining new, innovative, energy efficient technologies. In addition, she points out that the development of these alternative energy resources require significant capital investments that may have to be sustained over an extended period of time in potentially difficult economic contexts.

In the aftermath of the food price shocks of 2007 and 2008 and the estimated one billion people living with chronic hunger in 2009, food security was seen as a significant challenge for global policy makers and global institutions. Ben Shepherd argues for international co-operation on food security and examines bilateral deals over food producing resources as a vehicle of cooperation. He takes a critical look at the policy of agricultural "land grabbing" pursued by developed countries in sections of the developing world with the purpose of exporting food grown there back to their domestic markets to secure future food supplies. While these deals could be justified as providing an opportunity for investment and employment in the developing countries, it has been increasingly seen as a "modern form of colonialism". Compounding the concern is the absence of any international frameworks for ensuring that these bilateral arrangements would deliver desirable outcomes for the developing country parties. The potential for exploitation of developing countries was recognized and a 2009 Food and Agriculture Organization study offered several policy recommendations for investors and the host states. The report did not propose a more effective multilateral form of protection for the exploited parties and failed to emphasize the value of cooperative approaches in addressing food insecurity generated by these deals. The chapter examines a selection of the "land-grab" deals, analyzing the drivers and impacts for both the investing country and the developing country host. The paper assesses possible alternative, co-operative, regional and intra-regional frameworks for maximizing protection of the weaker parties in such deals while enhancing the long term food security of the states involved.

Water use and access has historically been an area of conflict when equitable distribution is not present. Shared water resources between multiple states are more likely to experience periods of tension. Fábio Albergaria de Queiroz's case study of the Prata Basin examines hydropolitics, security and cooperation at the South America Regional Security Complex. The author explores whether the case of La Prata, a dependence on shared water systems – both surface and groundwater – is of such a strategic nature that it drives inter-state relations of potential cooperation (amity) and/or competition (enmity) in a discernable manner. Noting that water is a vital resource with multiple uses that are essential to the stability of any region in the world, Queiroz argues that this is even more important for countries that share the resources of an international river basin. The concepts of securitization, securitizing actors, referent objects, security complex and the related concept of hydropolitical security complex as a distinct form of security complex where water is a salient feature of regional political dynamics are used to determine the impact of shared uses of the La Prata basin' resources by Argentina, Brazil and Paraguay. The idea of security as a relational phenomenon where hydropolitics issues are interpreted as an existential threat to a referent object is applied to this case study of a regional security agenda.

Daniel S. Papp's concluding chapter recognizes that the challenges of a globalizing world have necessitated a new and comprehensive approach towards the understanding of security which differentiates it from traditional ways of thinking about security. Reviewing a number of the challenges and threats presented in the previous chapters the author embellishes the observations and provides insight into additional threat sources including desertification and cyber security. He concludes that the new and broadly delineated security environment is dependent on solutions that include experimental and cooperative approaches if its frontiers are to be effectively defended.

The contributors to this work provide valuable insights into energy issues, as well as address the dilemmas faced by policy makers as they contemplate critical resources and traditional concerns of equitable and effective food and water access and distribution. Innovative regional and community specific policies, programs, and solutions are explored by several of the authors in their analyses. It is hoped that these thought provoking and informative studies will greatly enhance our understanding of a critical area of concern for the twenty-first century.

I am grateful to the contributors for their dedication and enthusiastic cooperation in making this volume possible. My thanks also go to Kirstin Howgate, Margaret Younger, and the Ashgate publishing team whose patience, encouragement, and guidance was crucial to the completion of this volume.

Chapter 1
Energy Security and the Revival of Geopolitics: The Russian Case

Mikhail A. Molchanov

Introduction

Before the world oil shock of 1973, the phrase "politics of energy" would connote about as much meaning as "politics of transportation" or "politics of frozen orange juice trade." Certainly, there was some politics to be found around energy production and trade, but that was clearly secondary to the politicians' more traditional preoccupations, such as national security, foreign policy, wealth creation or politics of alliances. Nowadays, none of these traditional preoccupations can be disengaged from the question of energy security. As the peak in global oil production nears, securing reliable sources of energy for domestic industrial and household consumption becomes a central theme in the discussions of national security and economic diplomacy, long-term developmental plans and foreign policy strategies. The times when "energy" was more associated with the word "policy" and only occasionally – with "politics" – are now long gone and forgotten.

In this chapter, the issue of energy security is being reviewed from a standpoint of geopolitics. The contention is that geopolitics is inescapably related to the question of energy security for a simple reason: the contradiction between the universal character of energy demand and the nation-state's exclusive right of management of natural resources located within its realm. The chapter starts with a discussion of the earlier historical antecedents of geopolitics of natural resources and the imperialist struggle for oil at the beginning of the twentieth century. It follows with an overview of the oil crises of the 1990s, which ended the era of the oligopolistic domination of the market by a few western companies, and instituted the second oil regime, much more dependent on stability of the producing countries. The discussion then moves to the impact of globalization and the return of geopolitics that followed mass disillusionment with the neoliberal project of the 1990s. The chapter ends with a case study that focuses on present struggles over the control of energy resources in Central Asia, Russia and Eurasia. As Russia reasserts controls over its energy production and trade, its energy policies and foreign economic relations take center stage in the discussions of energy security in both regional and global contexts.

The Rise of Energy Security

Energy security considerations lie at the intersection of the traditional security concerns and environmental problematique brought, on the one hand, by global climate change and, on the other, global resource depletion. If international relations realism determines relative standing and the very survival of the states by their material capabilities and the ability to enact them in pursuit of the power aggrandizement strategies, resources count, while the lack of resources betrays strategic vulnerability and endangers the state's very existence. However, what counts as strategic resources changes as human society evolves. Until the advent of the industrial revolution, men were arguably the ruler's main strategic resource. Horses and ships – essential vehicles for projection of armed forces – came close after. With all the energy needed having been supplied by readily available, replenishable, and seemingly inexhaustible sources, like manpower, horsepower, the sun, the wind or the water flow, the problem of energy security could not arise.

This all changed when new technologies of warfare demanded massive investments of non-replenishable resources, such as coal, oil, and, from the mid-twentieth century on, fissile materials. Vast energy inputs that modern war required made energy security into a subsystem of a modern national security complex as soon as western navies introduced steamships into the core of their striking forces. World War I, the first comprehensive industrialized warfare in the history of human civilization placed the question of energy security at the forefront of the national defense and international security agendas. Without a doubt, most early conceptualizations of the problem linked it to the necessity of adequate fuel supplies for the military forces.

Parallel to that, industrial economy's appetite for energy grew, making the nation's ability to provide it relatively cheaply to domestic businesses into a *sine qua non* of both domestic economic performance and comparative international standing vis-à-vis other industrialized nations of the world. Energy-intense economy appeared, for all intents and purposes, synonymous with economic progress and superiority, so that no less an authority on political and economic alternatives to capitalism than the founder of the Soviet state Vladimir Ilyich Lenin (1920) chose to describe his vision of a future communist society, strikingly, as "Soviet power plus the electrification of the whole country."

Russian Bolsheviks were certainly not uniquely prescient to understand the key significance of energy for economic development of a nation and to emphasize, therefore, the link between energy security and economic security writ large. While World War I was still raging, James Fairgrieve's musings on geography and world power had prompted a conclusion that history must be seen as "the story of man's increasing ability to control energy" (1915: 3–4). In the same book, Fairgrieve predicted that the gravity poles of world politics would soon start shifting from the energy-poor to energy-rich areas of the globe. The pioneers of "anthropogeography," such as German Friedrich Ratzel and American Ellen Churchill Semple, justified colonialism and neocolonial exploits of the West by the necessity to get access to natural resources that were in short supply back home. Colonization was described as "an outlet for national as well as commercial expansion," while prosperity of "both race and trade" was made contingent on "occupation of larger tracts of territory and the development of local resources as a basis for exchanges" (Semple 1911: 253).

The energy security discourse emerged spontaneously in the neocolonial pursuit of far-away sources of energy and raw materials. Those were increasingly required to power military machines of both old empires in Europe and the new, growing empires in North America and Japan. The emergence of mass-production automobile industry after 1900, the conversion of navies from coal to fuel oil and the advent of military aviation in World War I increased the world petroleum demand and pushed American, British and European companies to intensify the search for oil fields abroad. The advent of machine-based warfare gave new significance to energy security, leading Hans Niemetz to suggest that "land bases, oil, unified weather and news reporting are all that Germany needs" to develop an air force capable of nullifying British naval superiority (Gyorgy 1943: 359). Armed forces' dependence on oil increased exponentially since World War II, which had itself demonstrated, with all clarity, that "the question of adequate oil supplies has become a dominating factor in grand strategy and logistics" (Brooks 1944: 83). In the post-war period, it was observed that "the qualification of a great power today depends largely upon its minerals, above all coal, iron ore and oil" (Falls 1948: 5).

Geopolitics of Natural Resources

Today's debate on energy security has had its predecessors. Early geopolitical thinkers occasionally mention oil as a resource or in conjunction with the development of transportation routes which "might offer an opportunity" for this or that nation "to advance" to the known location of oil sources (Spykman and Rollins 1939: 408). In pre-World War II Germany, these scholars helped develop "policies of economic expansion which the exporters translated into practice" (Schnitzer 1955: 418). Emerging conceptualizations of energy security were often couched in geopolitical terms. Geopolitics remains an essential background for examinations of energy security for two main reasons: first, because both address the problem of resource determination of political agency, and second, because both assess relative power balances in the world system of states.

Considerations of energy security are those that relate to the state's ability to ensure that it has enough energy at its disposal to discharge its military and civilian functions effectively. The effective discharge of the state's military and defense functions is demonstrated by its success in repelling aggressors and cultivating allies. The state's domestic effectiveness is gauged by economic development and comparative social wellbeing of a nation. In both cases, domestic stability and international security of a nation is assessed through a comparison to other nations. Thus, energy security, while clearly related to such factors as the nation's natural resource endowments, geographic location and natural environment at large, should be approached as essentially a political category and a relational quality of international relations: whether one nation is "secure" or not in terms of energy inputs at its disposal is always determined by the concrete configuration of forces in the international arena.

Geopolitics examines "how geographical factors, including territory, population, strategic location, and natural resource endowments, as modified by economics and technology, affect the relations between states and the struggle for world domination" (Foster 2006: 1). The central precept of geopolitics has been traditionally the idea of political control and domination of physically defined space. Geopolitical analysis focuses primarily on spatial aspects of power distribution in the world and on geography as a resource in, or an impediment to, power accumulation and dispensation in international relations. Since power is largely a function of the state's material capabilities – its economic and technological reach, the size and equipment of its armed forces and the extent of its command of the material world generally speaking – studies of international politics cannot skirt the question of material endowments: demographic potentials and natural resources, energy first and foremost. Thus, the question of energy security intrinsically points at the question of the nation's relative power, or the lack thereof. If domestic energy resources are wanting, the nation must compensate with other instruments of influence at its disposal.

From its first appearance in Rudolph Kjellen's writings, geopolitics was concerned with the idea of national self-sufficiency (autarky) in natural resources. Karl Haushofer and his school borrowed the idea of self-sufficiency from Kjellen and advocated ready access to the sources of energy abroad as an important national objective for Germany. While Halford Mackinder was more concerned with global land and sea configurations as preconditions for power projection and influence, Haushofer applied organic concepts of the state and the idea of a *Lebensraum* ("living space") to validate imperialist conquests of resource-rich territories. German geopolitics differed from its British counterpart in its emphasis on what Haushofer called "the investigation of life-circumstances within a state and between states in their 'relationships of space'" (Weigert 1941: 10). This had important implications for the understanding of both power distribution in international relations and the strategies of resource acquisition. Thus, geopolitics could also be defined as a "global scheme of political strategy" (Strausz-Hupé 1972: vii).

For Mackinder, the world's center of power, the geographical "pivot of history," was to be found in the Eurasian Heartland – the core area of what he called the World Island. The Heartland's self-sufficiency in natural resources, in addition to its sheer size, strategic location, and the ability of the dominant land power to "close" navigable sea routes to its maritime competitors enabled Mackinder's famous dictum: "Who rules the Heartland commands the World-Island: Who rules the World-Island commands the World" (Mackinder 1919: 194). Since western ("sea-power") security relied on the security of sea lanes, it needed ready access to, if not control of, the Eurasian Heartland and its riches.

For Fairgrieve, the center of gravitation in international relations was to be determined by the location of the world's strategic sources of energy. Fairgrieve suggested that man's ability to control energy was subject to geographic limitations of a place (location of human habitat) and physical distribution of energy carriers around the globe. He wrote of a future where, with the exhaustion of coal and oil production in the North, the energy-rich regions of the global South were coming about to "keep or gain an importance at the expense of others not so fortunate" (Fairgrieve 1915: 334).

While both Fairgrieve and Mackinder's prescriptions to policy makers were mostly adaptive in nature, Haushofer took a more activist approach, thus harking back to Ratzel and Semple. For Haushofer, the idea of *Lebensraum* mandated nation's efforts to gain new space and resources to be used by growing population. As different nations grew at varying speeds, the resulting changes in international power structure dictated "the duty of the stronger state to expand at the cost of the weaker" (Gray and Sloan 1999: 226). Haushofer's theory echoed in Hitler's speeches, which prominently featured the notions of Lebensraum and self-sufficiency, as well as the idea of the state's duty to expand as needed to get the necessary resources for the nation. The quest for resources formed an integral part of German economic expansion strategy and informed Hitler's fateful decision to advance toward the oil-rich lower Volga and Northern Caspian areas.

Starting as a discourse on physical space uses in military strategy, geopolitics was soon transformed into the analysis of spatial preconditions for development. It was briefly discredited as a tool of aggressive German imperialism, but born again after World War II as evaluation of political risk factors in the development of natural resources. World capitalist integration and internationalization of commodity production have reinstated geopolitics as a respected subfield of international relations theory, which borders on international security studies, strategy, and international political economy. The energy security narrative bridges geopolitics and geoeconomics, while the state itself gets reframed as a geoeconomic agent par excellence (Cowen and Smith 2009). However, this new role has also had its historic precursors, as revealed in the special relationship between western governments and the oil industry since its inception.

The Scramble for Oil

The early players in the scramble for energy resources, such as the US-based Standard Oil (later Exxon/Mobil), the UK-based Anglo-Persian Oil Company (British Petroleum) and the Netherlands-based Royal Dutch/Shell, had successfully enlisted their respective governments into the struggle for concessions in Indonesia, Mesopotamia (Iraq), Latin America, Persia (Iran), Russia, Saudi Arabia, and Turkey. Imperial Britain took the lead, fighting, in Foreign Secretary Curzon's words, to keep control over "Turkestan, Afghanistan, Transcaspia, Persia" as the strategically located "pieces on a chessboard upon which is being played out a game for dominion of the world" (Kearns 2009: 215). These same areas – the Middle East, the North Caucasus, Iran, and Central Asia – were regions rich in oil, which Britain, concerned with the reliability of supply from the United States, wanted to secure for itself and deny its competitors (O'Hara and Heffernan 2006). The struggle between Britain and Russia led to the conclusion of two bilateral agreements (the Treaty of St. Petersburg, 1907, and the Constantinople Agreement, 1915) on the de-facto partition of Iran between the two powers.

The collapse of the Russian empire and the ensuing loss of the British oil concessions in the Caucasus moved Britain to invade both Russia (March 1918) and Azerbaijan (August 1918). The military campaign that began as an attempt to deny Germany the spoils of the Brest-Litovsk peace treaty with the Bolsheviks, oil in particular (Kearns 2009: 218), was crowned with the establishment of the British Oil Administration in Baku. At the time, Britain had already secured long-term concessions in Iran and established itself as the primary exporter of the Iranian oil. The question of energy security, which animated discussions of the British oil dependency on the United States since 1918, achieved apparent resolution in Viscount Harcourt's dictum that "every effort must be made now, and in the future, not only to develop existing oilfields in British territories or spheres of influence, but to acquire new fields that will be from the outset in British commercial hands and under British control" (O'Hara and Heffernan 2006: 64). This found full understanding of Maurice Hankey, the Secretary of the War Cabinet at the time, who argued that "the retention of the oil-bearing regions in Mesopotamia and Persia in British hands, as well as a proper strategic boundary to cover them, would appear to be a first-class British war aim" (Mejcher 1972: 385).

Equally, France sought a share of the Ottoman inheritance and, in a secret Sykes-Picot agreement with Britain, was promised control over Syria, Lebanon, southern Anatolia and northern Iraq, including parts of the oil-rich Mosul area. The French energy policy revolved around the ideas of a privileged partnership with the British, based on the two countries' joint administrative control over oil extraction and marketing (Fitzgerald 1994). Although not premised on the idea of exclusive territorial controls, the French policy nonetheless aimed to ward off attempts at penetration of the protected market by the third parties.

Americans joined the fray after the end of World War I and muscled their way into the Turkish Petroleum Company (TPC), then dominated by the British and the French. To counterbalance Europeans' insistence on spheres of influence, Americans advocated an open-door policy. The British had a head start, since Turkey had already granted a concession to the consortium representing Europe's companies, and Germany's war loss resulted in a transfer of its interests to the British. The British pipelines were to run across the French mandate territories, hence, the French could also bargain from a strong position. However, the American companies were not part of the deal, and the Department of State had to lobby on their behalf. The open door that Americans insisted upon was, in theory, open to all countries. In reality, it was meant to benefit American oil majors, and only they were able to make full use of the policy.

Out of these struggles, a new form of diplomacy was born. Oil started dominating international conferences: the 1920 San Remo agreement not only divided the Ottoman inheritance but served to accommodate the British and the French oil interests. Later, France was enticed to hand over control over the northern part of the Mosul province to the British in exchange for a 25 percent share in the Turkish Petroleum Company. The saga, however, did not end there, and the Lausanne Peace Conference in 1922–23 was, once again, absorbed by the Mosul question.

Similarly, the Soviet government had dangled prospects of oil concessions at the Genoa Conference in 1922. The Soviet delegation negotiated oil deals with the Italian government and Royal Dutch Shell, and nearly succeeded in breaking the united front of western representatives who tried to get Moscow to pay for confiscations in the Caucasus. Although no actual concessions had been granted, as a result of the conference, oil dominated Genoa discussions in early May 1922 and rumours of the Shell's success with the Soviets caused quite a stir, making the conference's atmosphere rather "greasy" (Petracchi 2002: 168).

Security of oil trade between the two world wars was discussed in terms of the state intervention or non-intervention in the "normal" functioning of the markets. The British policy of active state tutelage of the country's interests overseas stood in contrast with the American "pluralist" (laissez-faire) approach. However, both argued for the sanctity of the national companies' investments abroad. Meanwhile, the rise of nationalism and socialism world-wide had ushered in a period of changes that subjected foreign investments to political risk. Energy interests evolved from preoccupation with production rights in protectorates to fixation on the previously unseen phenomenon of nationalization – in Russia (1918–1922), Bolivia (1937), and Mexico (1938) – the situations that, in the US view, constituted "the chief abnormal elements in the world's petroleum problem" (Bedford 1923: 99). Government interference with domestic oil industry was considered the second-worst problem.

World War II replaced the British domination of the world oil market with that of the United States of America. A major, albeit temporary, change in US energy policy preceded that development. In 1942, President Roosevelt established the Petroleum Administration for War (PAW), vested with the task to coordinate oil production and refinement for America's war effort. PAW supplied US armed forces and, to a large extent, armed forces of the Allies, with 2.7 million tons of petroleum products shipped to the USSR alone. While PAW had been highly successful in meeting the war requirements, unprecedented levels of production ignited fears that US domestic oil supplies would not last long. The Roosevelt administration attempted negotiating an oil agreement with the British and developed plans for the government-owned Petroleum Reserves Corporation that would purchase all of the concession rights in Saudi Arabia from the Arabian

American Oil Company (ARAMCO). American geostrategists insisted that the US "would have to choose its current friends with an eye to their raw materials":

> Should the Near East oil lands, for example, be indicated by the geopolitical experts as of paramount importance, then American diplomacy, treaties, trade agreements, and the like would be directed toward securing rights, promoting local good will, and matching other competitors in this area against each other. A dummy international organization could be useful, but the United States would have to maintain their voting majority in the holding company that pulls the strings (Flanders 1945: 581, 583).

However, the PAW-steered movement for the state's more active involvement in oil industry had come to naught because of strong opposition of the US business interests. Finally, the idea of a global private cartel prevailed, while the plans for direct state intervention in oil industry or interstate agreement with Great Britain over the division of concessions in the Middle East had both collapsed.

1973 and After

The first oil regime was based on political domination of the Middle East initially by Great Britain, and later, the Anglo-American condominium. In that regime, western oil majors collectively set prices and determined levels of production. Western governments maintained an arm's length approach to the oil industry business matters. Those were negotiated by the members of the capitalist cartel themselves, often with little or no participation of the host country's government. While clearly oligopolistic in nature, the first regime served public interests of the western states "by providing a secure source of plentiful and cheap oil" (Frank 1985: 586).

That security was undermined by a series of developments that had been presaged by the Suez crisis of 1956–57 and the Arab-Israeli Six Day War of 1967. Both events had caused supply disruptions peaking up at about 2 million barrels a day. However, they could not yet alter the global system of oil production and distribution, which remained defined by the "seven sisters" consortium consisting of the Standard Oil of New Jersey (Esso, then Exxon), Standard Oil of New York (Mobil), British Petroleum, Shell, Texaco, Gulf Oil, and Standard Oil of California (Socal).

That system started unraveling when Libya forced concessionaries to pay higher prices for its crude in 1970. Nationalization campaigns in Algeria and Iraq in 1971–72 and the rise of the Organization of Petroleum Exporting Countries (OPEC), together with politically inspired Arab oil embargos against the West, sped up the already ongoing dilution of power of the Anglo-American condominium. Now the producing nations were able to set oil prices and the rate of production. The second oil regime was characterized by the fragmented pattern of organization, interdependence between exporting and importing countries, the loss of the know-how monopoly by western transnationals and the entrance of smaller independent producers (Pradhan 2008: 73–81). The collapse of the regime controlled by the "seven sisters" meant not only massive transfer of power and wealth from the developed to the developing parts of the world, but also new challenges to energy security of the western nations.

Oil crises of the 1970s inspired new works on economic diplomacy and the role of the state in ensuring energy interests of a nation. Several books discussed the rise of the OPEC and the use of oil as a political weapon. Several more were concerned with foreign economic policies of the West and the articulation of the state-market relationship. In one of the latter, Krasner (1978) argued that the United States, as a "weak state," did not do enough to protect the interests of its oil companies and acted out of political, rather than economic interests. He noted that security of supply seemed to be only one of the foreign economic policy goals of the US establishment, and was subordinate to "broader," or "more general" objectives:

> In cases concerning the protection of foreign raw materials investments, American policy-makers have consistently placed general foreign policy aims above security of supply, and security of supply above more competitive markets (Krasner 1978: 148).

Willrich (1976a) advocated "energy independence for America," pointing out that a superpower must be largely self-sufficient in energy resources and control enough of energy reserves to shield its military capabilities from potential disruptions of supply. Willrich defined energy security as a twofold concept involving both military and economic aspects. While the first meant "the guarantee of sufficient energy supplies to permit a country to function during war," the second was based on "the assurance of adequate energy supplies to maintain the national economy at a normal level" (Willrich 1976b: 747). Creation of strategic oil reserves, development of allocation and rationing plans on a standby basis, reduction in consumption and diversification of supply away from oil were all offered as potential solutions. Other US strategists focused on international diversification of supply, special partnerships with individual oil producing countries (Conant and Gold 1977) and even military intervention in the Persian Gulf.

The Cold War had a profound impact on geopolitics of energy security. Political challenges to the West-dominated regimes were either couched in ideological terms from the start or spun off as indicative of global confrontation between capitalism and communism regardless of their true origins. The unraveling of the colonial system and the rise of postcolonial nationalism were interpreted as the result of subversive activities by the Soviets. Reassertion of national sovereignty over resources was read as a threat to their extraction. Renegotiation of the unfair deals imposed by western corporations could be decried as "virtual nationalization." When seen through the geopolitical prism of the Cold War thinking, even modest redistributional efforts affecting American interests abroad could be perceived as detrimental to "the core objectives of American central decision makers" (Krasner 1979: 91).

The Iranian Revolution of 1978 and the Soviet invasion of Afghanistan next year ignited fears of the impending loss of access to oil fields in the Middle East. The Iranian Revolution resulted in the largest ever supply disruption of 5.6 million barrels per day for a period of six months. This, in addition to the Soviets' advances in Southwest Asia, provoked an alarmist reaction in the White House. "The Soviet intervention in Afghanistan constitutes one part of a giant pincer movement designed to encircle Gulf oil reserves," went a typical commentary of the day (Phillips 1980: 6). The fact that the USSR was mostly concerned with the Afghan domestic power struggles and did not attempt moves of consequence to the oil industry in the region did not make any difference.

On January 23, 1980, President Jimmy Carter reacted to the events by asserting America's right to intervene with force, if necessary, to ensure uninterrupted energy supply from the Gulf. The speech, which was drafted by then national security adviser Zbigniew Brzezinski, proclaimed that "an attempt by any outside force to gain control of the Persian Gulf region" would be regarded "as an assault on the vital interests of the United States of America," and would have to be repelled "by any means necessary, including military force." Parallel to that, US strategists suggested that only a "hemispheric energy effort," i.e. maximization of supply from Canada, Mexico and Venezuela, could help curb US dependency on Gulf oil (Rutledge 2006: 46–9).

The Carter Doctrine proved a long-standing instrument of American foreign policy. Every US president since 1980 has reaffirmed the doctrine's basic principle and enhanced US rapid deployment forces in the Gulf (Klare 2009). Ronald Reagan had applied the doctrine at the height of the Iran-Iraq war to protect the US-flagged Kuwaiti tankers passing through the Strait of Hormuz. George H.W. Bush sought to defend US oil interests in Saudi Arabia and authorized Operation Desert Storm, which repelled Iraq's aggression against Kuwait. George W. Bush commanded a US invasion of Iraq in 2003 under the pretext of stopping Saddam Hussein's efforts to develop weapons of mass destruction; the real reason for the war was the US desire to establish a geopolitical bridgehead in the region and secure control over strategic energy resources (Klare 2004; McQuaig 2004; Bromley 2005). Finally, just as President Clinton's support of the sanctions and limited military pressure on Iraq (the "no-fly zones"), President Obama's insistence on "an over-the-horizon force that could prevent chaos in the wider region" illustrates the continuing relevance of the doctrine.

Western countries had also sought to counterbalance OPEC power collectively. These efforts led to the creation of the International Energy Agency (IEA) in the framework of the Organization for Economic Co-operation and Development (OECD). IEA's original mandate was to facilitate emergency preparedness through the establishment of oil reserve stocks in the participant countries and their release in times of need – the so-called "stockdraw." The 1974 Agreement on an International Energy Program (I.E.P.) made it obligatory for the signatories to hold oil stocks equivalent to at least 90 days of the previous year's net oil imports and to use the inventories to balance markets in the event of a supply disruption. In addition to the stockdraw itself, the Agreement advised such measures as demand restraint, increased domestic production, fuel switching, and sharing of available oil between member states, if necessary.

Globalization and the Revival of Geopolitics

The Cold War's end changed the ideological, political, and economic landscape of energy security. If the Iran-Iraq war in the 1980s could still be seen as a superpower confrontation by proxy, and a concern could still be raised over the USSR "exploiting fears of interruption of oil deliveries from the Persian Gulf" (CIA 1980: 3), the US invasion of Iraq in 1991 had ushered in a unipolar moment in world politics. It was the first interstate war that was fought specifically over oil and perhaps the first big war of the modern era that did not cause bitter divisions among major powers. The latter factor spoke to the "end of history" in Fukuyama's sense – ideological divides were passing into oblivion, while capitalist globalization was finally making its entrance as not only economic and technological, but also political reality.

With the collapse of the USSR and China's abandonment of the socialist idea in practice, if not in theory, the struggle for oil ceased to be guided by ideological reasoning. The ultimate demise of communism, the "rolling back" of the state and the apparent triumph of neoliberalism, the rise of global environmental concerns and the growth in Asian, and especially Chinese, energy demand have all contributed to an understanding of energy security as a shared, global problem. Even though concerns over strategic vulnerability of supply remained and were occasionally dramatized in various arenas, the superpower confrontation was thing of the past and so were, for the time being, conspiracy theories presenting the problem of energy security as a resultant of some clandestine plot for world domination. Throughout the 1990s, the threefold global challenge of oil dependency, resource competition and energy security evolved as principally economic in nature.

As the first decade of the new millennium unrolled, there was a moment of rather naïve optimism. American think tanks promised a decline of resource nationalism and rejoiced at global acceptance of private sector solutions. Globalization appeared synonymous with liberalization, i.e. "a process of removing officially imposed restrictions on movements of resources between countries in order to form an 'open' and 'borderless' world economy" (Scholte 2009: 95). Energy security seemed achievable via encouragement of business-friendly regimes and deregulation of the markets. America's withdrawal from the Kyoto Protocol indicated that its government saw global warming as little more than a pet issue blown out of proportion by the greens. The alarmist calls about resource depletion were heeded even less. The North-South disagreements over environmental policies looked more important than supply interruptions, and even terrorism appeared as a "temporary jeopardy" (CSIS 2000: 9).

This optimism evaporated after the September 11, 2001, terrorist attack on America, the Bush Doctrine and the wars in Afghanistan and Iraq, which many observers saw as guided in no small part by the US oil appetites. A political dimension of energy security was swiftly reintroduced to reflect the new environment characterized by the emergence of global terrorism, the unopposed ride of the USA as the world solo "hyperpower" and the rise of China as the US potential challenger. Security of oil supply is back, and center stage. Nationalization of oil and gas industries in Bolivia and Venezuela, expansion of the state control in Ecuador and Brazil, de-facto nationalization in Russia and the rise of the "national champions" on the African continent from Algeria to Angola have reversed the heralded decline of resource nationalism. If anything, it is on the rise.

The revival of energy geopolitics in the post-9/11 world has been preprogrammed by a series of events that ranged in scope from regional to global. First, the widespread reaction to two decades of neoliberalism led to its rejection in favor of more skeptical, if not more balanced, approaches in public policy. The 1997–99 global economic crisis, which started in east Asia and ended in Argentina, had dealt the idea of unregulated financial markets a big blow. Multiple market failures in 2007–2010 required government bailouts and application of stimulus packages. All of a sudden, governments were back in vogue, and speaking the language of national interests. The assurance of adequate energy supplies has reappeared among the government's core preoccupations.

Second, the trend toward a more activist government evolved in parallel with growing dissatisfaction with America's global leadership. The later sentiment could only grow worse in the wake of the Bush-Cheney essentially unlawful intervention in Iraq. Blatant manifestations of the US "go it alone" approach to international politics rendered the common good rhetoric of neoliberal institutionalists shallow. National reassertion and rejection of America's unilateralism propelled the left turn in Latin America, the rise of statist neo-authoritarianism in China and Russia, Iran's nuclear challenge to the West and intensification of resource nationalism in Africa. Resource politics acquired features of a zero-sum game guided by such principles as "what is mine is not yours" and "one side's gain is another side's loss."

Resource geopolitics of the twenty-first century is dominated by the three political risk factors:

a. (a) access to resources,
b. (b) security of supply, and
c. (c) long-term impact on the environment.

Of these three, only the last one can be considered a genuinely new policy driver. Against the background of global climate change, freedom from oil becomes a key dimension of energy security. Even so, the laudable goal to develop cleaner energy sources is often conflated with geopolitical considerations of achieving security from external man-made threats. As President Bush said in his 2006 State of the Union address, "America is addicted to oil, which is often imported from unstable parts of the world." Instability in oil-supplying regions of the developing world jeopardizes stability of the developed world. To overcome this strategic vulnerability, America has to leave its petroleum-based economy behind, thus breaking dependence on oil-exporting regimes in the Middle East.

Meanwhile, resource competition remains a constant feature of international relations. Both energy excesses of Western consumer-driven economy and the unstoppable migration of industries to the East breed fears of "resource wars" (Klare 2001) that the political economy of global climate change can only intensify. These wars do not need to bring great powers into a face-to-face confrontation; at least that possibility may still be some time away. However, the obstructable nature of oil production and transportation makes it an easy target for terrorism and invites violent confrontations over property rights and rents redistribution (Le Billon 2005). The struggle for oil rents often fuels separatist conflicts. It may also ignite civil wars and coups d'etat, as well as terrorist activities ranging from kidnapping of oil workers for ransom (Columbia, Nigeria) to blowing up pipelines and infrastructure (Iraq, Kurdish rebels in Turkey, Gaddafi forces in the Libyan civil war). It breeds corruption and violent opposition to the corrupt elites from Nigeria to North Caucasus.

The threat presented by non-state actors and transnational terrorist networks is a new feature brought in by globalization, regulatory and public policy failures in the North and postcolonial state failures in the South. Even so, major supply interruptions so far have been commonly caused by the exporting nations and governments (nationalization, contract renegotiation, regulation and policy changes) and transit countries (pricing disputes, contract renegotiation), rather than anti-government rebels or rent-seeking terrorists. In addition to that, supply interruptions in the international gas market are not infrequently caused by the importer, which may opt to retaliate against price increase by the exporter, or refuse to accept the full volume of the earlier contracted gas because of changed economy or downstream losses (Hayes and Victor 2006: 332–8).

Although geopolitical contestations remain primarily the purview of national governments, energy firms play a substantial role in shaping both home and host countries' policies. Transnational corporations, headquartered predominantly in oil importing countries, became de-facto kingmakers for a number of oil- and gas-producing countries in the world's periphery and semi-periphery. These corporations' goals – free access to resources and security of supply – often coincide with geopolitical objectives of their home countries. Ultimately, both goals can only be guaranteed with the preponderant military presence that makes host countries and populations hostages to the ambitions of a hegemonic power overseas. Thus, geoeconomics of resource extraction becomes conflated with the old-fashioned imperialist geopolitics of a "grand chessboard" (Brzezinski 1997).

At the same time, as the 2010 BP oil spill in the Gulf of Mexico demonstrated, the governments of the leading capitalist nations may come at loggerheads with the transnational corporations that form a vital part of their own energy infrastructure. In John Ruggie's (2010) words, the governance gaps that separate preventative measures from the remedial ones and the state's duty to protect from the corporate responsibility to respect human rights "provide the permissive environment for wrongful acts by companies of all kinds without adequate sanctioning or reparation." Environmental concerns and consumer prices are two areas of concern where big oil interests are clearly at odds with those of general public. To give but one example, while overseas crises drive retail prices up and therefore do not answer consumers' preferences, energy giants may well stand to benefit from the windfall. A systemic discrepancy between the national interest and self-serving business interests creates conflicting geopolitical impulses on foreign policy making and implementation in advanced capitalist states.

Energy Politics in Russia and Eurasia

Ostensibly isolated from the world market economy, the Union of Soviet Socialist Republics engaged in its own version of international energy politics as early as 1970s. Under Leonid Brezhnev and Alexei Kosygin, gas trade with Western Europe became a tool of Soviet diplomacy that, among other things, pushed Willy Brandt's government into the adoption of radically innovative "ostpolitik." The "gas-for-pipes" deal ensured a steady supply of German-made pipes for the growing Soviet gas network that traversed countries of the Eastern bloc, cementing their loyalty to the regime in Moscow. The Siberian gas then continued its journey all the way to the West German customer.

The use of preferential pricing in Soviet oil and gas trade with other members of the Council for Mutual Economic Assistance (COMECON) and friendly regimes worldwide amounted to both implicit and explicit energy subsidies. Cuba and Vietnam in particular were among the primary recipients of Soviet largesse, which played a not negligible role in helping them to overcome dire economic consequences of the US-imposed blockade. Without exaggeration, Soviet energy supplies greased the wheels of the overall production process of the COMECON member states and, in many instances, kept the otherwise inefficient and internationally uncompetitive industries in Eastern Europe afloat. The true scope and significance of these energy subsidies became apparent only in the wake of the collapse of communism in Central and Eastern Europe and Eurasia, as industries went down and new entrants into the world capitalist club had to grapple with harsh reality of the world market prices on their own.

While energy subsidies had ended with the end of communism, oil politics had soon reappeared in both international and domestic struggles for power and influence. Privatization of the energy sector in Russia, Kazakhstan and Azerbaijan had divested states of valuable resources. Intense jockeying for the national slices of one of the largest energy pies on the planet had changed economics and transformed the security calculus of these states. The chaotic sell-off of national resources had caused a global scramble for the control of oil and gas fields in Eurasia. No government of a G7 nation could stay aside and not get involved. Oil majors had all showed up in Russia and the Caspian littoral states of the former Soviet Union. Not a single transnational could let it pass on the opportunity. Privatizing governments had soon found themselves outflanked by business players.

In addition to business executives, the US government officials, too, came in force. The US-Azerbaijan Chamber of Commerce had boasted names of Dick Cheney, Zbigniew Brzezinski and Henry Kissinger, along with Brent Scowcroft and James Baker, on its "Honorary Council of Advisors," while Richard Armitage was a member of its Board of Directors and Richard Perle sat on the Board of Trustees (Rutledge 2006: 108). Politicians were followed by geostrategists, intelligence, and military planners, who worked in full awareness of western business interests in the area. Finally, the US war in Afghanistan added extra clout to American oil majors negotiating productions rights and transportation agreements with the governments of Transcaucasian and Central Asian states (Johnson 2004: 172–3)

The Russian Federation occupies centre stage in most accounts touching upon the politics of energy in the region. Energy trade preserved Russia's ties with the newly independent states of the former Soviet Union. Moscow has also used its control over the pipeline routes to the West to obtain political and economic concessions from Ukraine, Belarus, Turkmenistan, and Kazakhstan in particular. More than one author has noted that Russia seemed to be using its energy resources as a tool of foreign policy toward the so-called near abroad (e.g., Nygren 2008; Stent 2008). Studies of continental energy security extend the same logic of analysis onto Russia's relations with the EU states.

If extracting political concessions by means of energy trade is something that many note in Russia's behavior, there is much less agreement on who, exactly, is calling the shots. Undoubtedly, the Kremlin is very much in control. But what is the role of the energy companies? While Smolansky (1999) does not differentiate between the state and the energy firms that do its bidding, a more cautious position recognizes certain symmetry, but not a full coincidence, between them (Balmaceda 1998). Others see energy producers as relatively independent (Adams 2002). Both Stern (2005; 2006) and Goldthau (2008) dispute the use of energy trade as a foreign policy weapon and argue that Gazprom's behavior in particular speaks to a desire to maximize profit rather than advance foreign policy designs.

The debate on Russia's role in Central Asia brings back the memories of Mackinder and Haushofer. Thanks to Brzezinski (1997), energy politics in the region became presented as a new "Great Game" for the twenty-first century. Just as the classic Great Game, the present struggle over control of the Caspian Sea energy resources involves world powers that are driven into a geopolitical competition over prized territories in Central Asia and beyond. The classic competition involved Great Britain and the Russian Empire. The present one pitches Russia against the United States of America. In addition, Russia competes with Great Britain and other European powers; Russia and China use the Shanghai Cooperation Organization as a regional shaker; smaller states join forces to stave off advances of outside powers; and local pretenders at regional leadership try to outflank one another.

Geopolitical significance of the new Great Game may well be blown out of proportion (Lieven 1999/2000). Nonetheless, stability in the Transcaspian region affects stability of its neighbors, and must be assured if an effective international regime for the development of region's energy resources is ever to be put in place (Molchanov and Yevdokimov 2004). Regional stability answers Russia's economic and foreign policy interests, which both zero in on the necessity to safeguard national independence and facilitate development.

This analysis sees Russia's energy politics as mostly defensive, and rarely offensive, in nature. Moreover, it is often merely reactive. A rather famous episode of Yeltsin's using a "tap weapon" (Nygren 2008) to push Ukraine's denuclearization transpired in the context of trilateral negotiations with the USA and served American designs toward Ukraine no less than Russia's own preferences (Molchanov 2002: 233–4). Similarly, disruptions of the gas flow through Ukraine in early 2006 and 2009 followed 15 years of non-payment of $1.4–$2 billion gas debt that Ukraine owed to Russia and routine theft of gas flowing through Ukraine en route to Europe. Far from vindicating Dick Cheney's rhetoric of "energy blackmail," these temporary disruptions of supply may, in fact, speak of Russia's determination to secure gas supplies to the EU customers.

Energy diplomacy in East Asia revolves around oil and gas pipelines. Originally labeled the "Angarsk-Daqing Line," the China-Russia oil pipeline project was first transformed into the Eastern Siberia-Pacific Ocean (ESPO) pipeline to the port of Nakhodka, where it would serve primarily Japan. Upon Beijing's protestation, a compromise emerged to build a branch connector to China, with two-thirds of oil going to Japan and the remainder to the PRC. The new routing promised to maximize Russia's benefits because of buyers' competition (Buszynski 2006). In 2008–2009, agreements were signed on expedited construction of the Chinese trunk and long-term (2011–2030) supplies of Russia's oil to China in exchange of $25 billion in loans to the state-controlled Rosneft and Transneft corporations.

With the completion of the first stage of the ESPO pipeline and the launch of a new Russian oil port in Kozmino in late 2009, oil started being shipped to East Asian markets. Journalists noted the symbolic significance of choosing Hong Kong for the ESPO oil's first shipment overseas. East Siberian crude is hereby used to give a new boost to the Russo-Chinese strategic partnership. The completion and filling of the ESPO direct spur to China (Skovorodino-Daqing) promises to change the geopolitical landscape dramatically. Because of sheer quantity of the Siberian reserves brought to the market, Russia may develop a new and significant leverage in negotiations with potential buyers in Europe and East Asia alike.

Both China and Japan are heavily involved in Russia's Far East energy system, and will be investing more. While the April 2009 agreement secures China's deeper engagement, Japanese interests remain well represented in Sakhalin I and Sakhalin II oil and gas projects. South Korea showed interest in ESPO-related spot deals and was promised joint projects to develop West Kamchatka offshore gas deposits. The construction of a new gas pipeline alongside the ESPO's route will bring new substance to economic cooperation in the region. It has been estimated that the Asia-Pacific markets, which absorbed 3 percent of Russia's oil exports in 2003, may absorb up to 30 percent in 2020. Gas exports to Asia Pacific will correspondingly grow from 5 percent in 2006 to 25 percent by 2020 (*Vladivostok Times*, April 28, 2006).

Thus, the Putin-Medvedev energy diplomacy in East Asia has been largely successful. It sought to avoid tying energy exports to one market only, and it has succeeded in achieving a balance between China and Japan. Moreover, with ESPO's launch, South Korea established itself as a major importer of the Siberian crude. Russian oil satisfied 7.3 percent of the country's demand in May 2010. In addition to that, South Korean companies were promised a stake in the West Kamchatka gas fields. With more Asian interests coming on board, Moscow is getting new instruments of influence over the European Union customers. Russian market presence has grown dramatically in the Pacific United States, where imports of Russian oil went from zero to an estimated 100,000 barrels a day by July 2010. East Siberian pipelines boost Moscow's positions in negotiations with the West on a range of issues from oil prices to loans to foreign direct investments to importation of new technologies.

However, Russia was less successful in the Transcaspian. Here, Azerbaijan and Kazakhstan aggressively compete for foreign direct investment. While the Baku-Novorossiisk and the Caspian Pipeline Consortium (CPC) pipelines are controlled by Russia, the Baku-Supsa and the Baku-Tbilisi-Ceyhan (BTC) lines fully avoid Russian territory. The same idea – to wrench the prospective transit routes out of a zone controlled by Moscow – lies behind the currently executed plan for the Nabucco gas pipeline, which will serve to reduce Europe's dependence on Russian gas. The Kazakhstan-China oil pipeline, opened in 2006, undermines Russia's near-monopoly of the land-based oil exports to East Asia, while the Turkmenistan-China gas pipeline, opened in 2009, frees Central Asian exporters from dependence on Russia's transit networks.

Politically and economically, the pro-western stance of governments in Azerbaijan and Kazakhstan proved a major obstacle to Russia's reasserting itself in the region. The US and British governments' determination to undercut Moscow's influence provided a yet bigger hurdle. Geographic remoteness of the region and technical limitations of the existing pipeline infrastructure further restrict the space for maneuver. Transcaspian energy politics is therefore defined by competing visions of export development, the western governments and TNCs' direct involvement in the region, and proliferation of transportation networks that diversify supply and undercut potential monopoly or monopsony powers of any single agent.

Putin proposed to increase the supply of Russian hydrocarbons to China via the pipeline network running through Kazakhstan. Russia's interest in filling the pipelines built by China is no less geopolitical than it is commercial. If the bulk of the Kazakh oil flows east, the geopolitical significance of the US-backed Baku-Tbilisi-Ceyhan pipeline will be undercut. Moreover, Russian oil pumped through the Central Asian network will further strengthen Sino-Russian political and economies ties. In 2009, Russia exported slightly more than 30,000 barrels per day via the Kazakhstan-China pipeline.

Russia's Energy Strategy represents oil and gas exports as a preferred instrument for the conduct of the country's domestic and foreign policies. Russia chose energy security as the topic for the G8 summit it hosted in 2006. However, its reading of energy security differs from the one preferred in the West. Russia understands energy security as the necessity to deal with the new geopolitical challenges born out of the global competition for resources. After all, it was the USA that initiated a new Great Game in Central Asia. Western interests tried to undermine Russia's role at the core of the Eurasian pipeline networks. The BTC and the Nabucco were both devised to that end. For the West, energy security is about diversification of supply. For Russia, it means stability and risk sharing. Hence, western geopolitics of energy may run at odds with Russia's approach to energy security as "a fair distribution of the risks among energy resource producers, transit service providers and consumers" (Putin 2006).

Conclusion

"Geopolitical" characterizations of Russian foreign economic policy would sound hollow if applied to a smaller country. However, when used to analyze a country that spans nine time zones and occupies close to one-eighth of the world's landmass, the one with truly continental reach and cross-border trade in Europe, Asia, and America alike, such characterizations are to be expected.

Because of its sheer size and the role as a continental bridge that links the three centers of global capitalism together into the single industrialized North, Russia would be geopolitically important even as a stretch of barren, desolate land devoid of any usable resources. Since it is, in fact, an energy superpower that has overcome Saudi Arabia in oil production and exports and holds more than one quarter of the world's gas reserves, it appears a crucial factor in any consideration of energy security.

Geopolitical significance of Russia's energy trade is fourfold. First, by being able to provide considerable energy supply to both old (EU, Japan, USA) and new (China, South Korea, Taiwan, potentially India) industrialists, Russia is positioning itself as a safety valve against potential disruptions of supply from the less stable areas of the world. Second, by being able to switch between the European and Asia-Pacific customers, Russia acquires not only a stronger position for energy sales, but also an instrument of geopolitical leverage over both groups of countries. Third, as an owner of a continental network of oil and gas pipelines, it can offer downstream and intermediary services to its Transcaspian competitors. Its buyer's power in the region, although no longer monopsonic, is still formidable. Finally, it is able to influence countries reliant on energy transit fees. Russia's ability to choose between transportation routes at its disposal means that it may either reward or punish the owners of transit networks that are dependent on the steady flow of Russian oil and gas exports.

The Russian case is illustrative of the problems that emerge on intersection of geopolitics and energy security. The contradiction between the universal character of energy demand and the nation-state's exclusive claim to its natural resources is being manifested in the tensions that arose out of Russia's bringing most of its oil and gas production under the state control and the state's attempts to ensure the best possible terms of trade for the national champion firms, such as Gazprom. The "pipeline wars" in the Transcaspian resurrect geopolitical musings on the strategic role of geographic space and transportation routes as means to international power projection. Essentially, geopolitical idea of resource determination of political agency has found a solid proof in Russia's rise from the impoverished and demoralized write-off of its former Soviet self to the newfound assertiveness that reaches well beyond energy into the traditional domains of international security and great power politics.

Since energy security is about our ability to control energy, and energy reserves are subject to uneven spatial distribution in the world of states, the state that controls substantially more energy than it needs and can assert its rights over resources will always be able to punch above its weight. Russia is one example of this rule, but so are, also, Saudi Arabia, Iran, Venezuela, or Libya. Russia is just bigger than the others and, hence, may be playing geopolitical games over resources under its jurisdiction somewhat longer than the rest.

References

Adams, J.S. (2002). Russia's gas diplomacy. *Problems of Post-Communism* 49(3), 14–22.

Balmaceda, M.M. (1998). Gas, oil and the linkages between domestic and foreign policies: The case of Ukraine. *Europe-Asia Studies* 50(2), 257–86.

Bedford, A.C. (1923). The world oil situation. *Foreign Affairs* 1(3), 96–107.

Bromley, S. (2005). The United States and the control of world oil. *Government & Opposition* 40(2), 225–55.

Brooks, B.T. (1944). *Peace, Plenty and Petroleum*. Lancaster, PA: The Jaques Catell.

Brzezinski, Z. (1997). *The Grand Chessboard: American Primacy and its Geostrategic Imperatives*. New York: BasicBooks.

Buszynski, L. (2006). Oil and territory in Putin's relations with China and Japan. *The Pacific Review* 19(3), 287–303.

CIA (Central Intelligence Agency). (1980, 24 December). *Special National Intelligence Estimate: Soviet Interests, Policies, and Prospects with Respect to the Iran–Iraq War*. SNIE 11-34/36.2-80.

Conant, M., and Gold, F. (1977). *The Geopolitics of Energy*. Washington, DC: US Government.

Cowen, D., and Smith, N. (2009). After geopolitics? From the geopolitical social to geoeconomics. *Antipode* 41(1), 22–48.

CSIS (The Center for Strategic and International Studies). (2000). *The Geopolitics of Energy into the 21st Century. Vol. 3: The Geopolitical Outlook, 2000–2020*. A report of the CSIS Strategic Energy Initiative. Project co-chairs, Sam Nunn, James R. Schlesinger; project director, Robert E. Ebel. Washington, DC: CSIS Press.

Fairgrieve, J. (1915). *Geography and World Power*. London: University of London Press.

Falls, C. (1948). Geography and war strategy. *The Geographical Journal* 112(1/3), 4–15.

Fitzgerald, E.P. (1994). France's Middle Eastern ambitions, the Sykes-Picot negotiations, and the oil fields of Mosul, 1915–1918. *The Journal of Modern History* 66(4), 697–725.

Flanders, D.P. (1945). Geopolitics and American post-war policy. *Political Science Quarterly* 60(4), 578–85.

Foster, J.B. (2006). The new geopolitics of empire. *Monthly Review* 57(8), 1–18.

Frank, L.P. (1985). The first oil regime. *World Politics* 37(4), 586–98.

Goldthau, A. (2008). Resurgent Russia? Rethinking Energy Inc. *Policy Review* 147, 53–63.

Gray, C.S., and Sloan, G.R. (eds). (1999). *Geopolitics, Geography, and Strategy*. London: Frank Cass.

Gyorgy, A. (1943). The geopolitics of war: Total war and geostrategy. *The Journal of Politics* 5(4), 347–62.

Hayes, M.H., and Victor, D.G. (2006). Politics, markets, and the shift to gas: Insights from the seven historical case studies. In D.G. Victor, A.M. Jaffe, and M.H. Hayes (eds), *Natural Gas and Geopolitics: From 1970 to 2040* (319–53). Cambridge; New York: Cambridge University Press.

Johnson, C.A. (2004). *The Sorrows of Empire: Militarism, Secrecy and the End of the Republic.* London: Verso.

Kearns, G. (2009). *Geopolitics and Empire: The Legacy of Halford Mackinder*. Oxford: Oxford University Press.

Klare, M.T. (2001). *Resource Wars*. New York: Metropolitan.

Klare, M.T. (2004). *Blood and Oil: The Dangers and Consequences of America's Growing Dependency on Imported Petroleum*. New York: Metropolitan.

Klare, M.T. (2009). Repudiate the Carter Doctrine. *Foreign Policy in Focus*, January 22. Retrieved September 29, 2011, from http://www.fpif.org/articles/repudiate_the_carter_doctrine.

Krasner, S. (1978). *Defending the National Interest: Raw Materials Investments and US Foreign Policy*. Princeton, NJ: Princeton University Press.

Krasner, S. (1979). A statist interpretation of American oil policy toward the Middle East. *Political Science Quarterly* 94(1), 77–96.

Le Billon, P. (2005). The geopolitical economy of "resource wars." In P. Le Billon (ed.), *The Geopolitics of Resource Wars: Resource Dependence, Governance and Violence* (1–28). London; New York: Frank Cass.

Lenin, V.I. (1920). Report on the work of the Council of People's Commissars. Eighth All-Russia Congress of Soviets. *Collected Works*, vol. 31. Retrieved September 28, 2011, from http://marxists.org/archive/lenin/works/1920/8th cong/ch02.htm.

Lieven, A. (1999/2000). The (not so) Great Game. *National Interest* 58, 69–80.

Mackinder, H. (1919). *Democratic Ideals and Reality*. New York: H. Holt.

McQuaig, L. (2004). *It's the Crude, Dude: War, Big Oil and the Fight for the Planet*. Toronto: Doubleday Canada.

Mejcher, H. (1972). Oil and British policy towards Mesopotamia, 1914–1918. *Middle Eastern Studies* 8(3), 377–91.

Molchanov, M.A. (2002). *Political Culture and National Identity in Russian-Ukrainian Relations*. College Station, TX: Texas A&M University Press.

Molchanov, M.A., and Yevdokimov, Y. (2004). Regime building as a prime mover of technological progress: The energy sector in the Central Asia-Caspian region. *Perspectives on Global Development & Technology* 3(4), 417–35.

Nygren, B. (2008). Putin's use of natural gas to reintegrate the CIS region. *Problems of Post-Communism* 55(4), 3–15.

O'Hara, S., and Heffernan, M. (2006). From geo-strategy to geo-economics: The "Heartland" and British imperialism before and after MacKinder. *Geopolitics* 11(1), 54–73.

Petracchi, G. (2002). Italy at the Genoa conference: Italian-Soviet commercial relations. In C. Fink, A. Frohn, and J. Heideking (eds), *Genoa, Rapallo, and European Reconstruction in 1922* (159–70). Cambridge: Cambridge University Press.

Phillips, J. (1980). The Soviet invasion of Afghanistan. *The Heritage Foundation Backgrounder* No. 108. Retrieved September 29, 2011, from http://s3.amazonaws.com/thf_media/1980/pdf/bg108.pdf.

Pradhan, S.R. (2008). *India, GCC and the Global Energy Regime: Exploring Interdependence and Outlook for Collaboration*. New Delhi: Academic Foundation.

Putin, V. (2006). Vladimir Putin's speech at Meeting with the G8 Energy Ministers. March 16, 2006, Moscow. Retrieved September 29, 2011, from http://www.g8.utoronto.ca/energy/energy_putin060316.html.

Ruggie, J. (2010). Operationalizing a business and human rights framework. Report of the Special Representative of the Secretary-General on the issue of human rights and transnational corporations and other business enterprises. United Nations Human Rights Council. *Policy Innovations*, April 23.

Retrieved September 29, 2011, from http://www.policyinnovations.org/ideas/policy_library/data/01577.

Rutledge, I. (2006). *Addicted to Oil: America's Relentless Drive for Energy Security*. New York: I. B. Tauris.

Schnitzer, E. (1955). German geopolitics revived. *The Journal of Politics* 17(3), 407–23.

Scholte, I.A. (2009). Defining globalization. In D. Greenaway (ed.), *Global Trade Policy 2008* (91–122). Oxford: Blackwell.

Semple, E.C. (1911). *Influences of Geographic Environment on the Basis of Ratzel's System of Anthropo-geography*. New York: Henry Holt.

Smolansky, O.M. (1999). Fuel, credit, and trade: Ukraine's economic dependence on Russia. *Problems of Post-Communism* 46(2), 49–58.

Spykman, N.J., and Rollins, A.A. (1939). Geographic objectives in foreign policy, I. *The American Political Science Review* 33(3), 391–410.

Stent, A.E. (2008). Restoration and revolution in Putin's foreign policy. *Europe-Asia Studies* 60(6), 1089–106.

Stern, J.P. (2005). *The Future of Russian Gas and Gazprom*. Oxford; New York: Oxford University Press.

Stern, J.P. (2006). Natural gas security problems in Europe: The Russian-Ukrainian crisis of 2006. *Asia-Pacific Review* 13(1), 32–58.

Strausz-Hupé, R. (1972). *Geopolitics: The Struggle for Space and Power*. Reprint edition. New York: Arno Press.

Weigert, H.W. (1941). *German Geopolitics*. New York; Toronto: Oxford University Press.

Willrich, M. (1976a). Energy independence for America. *International Affairs* 52(1), 53–66.

Willrich, M. (1976b). International energy issues and options. *Annual Review of Energy* 1, 743–72.

Chapter 2
Energy Transitions and International Security in the Twenty-first Century

Mert Bilgin

Introduction: Energy Security, Energy Transitions, and FAST Matrix

Energy defines the thermodynamic quantity needed by a system of mechanic, organic, electronic and of other structures to do work; work defined as mental and/or physical activities of all kinds for any reason (Grant 1888, Cook 1971, Andrews and Jelley 2007, Weissenbacher 2009). It is therefore the most essential need, which evokes the concept of security. Uninterrupted and affordable supplies of energy resources are the minimum conditions for energy security. Uninterrupted supply means the presence of energy for immediate use when ever, where ever, and as much as, it is needed. Amount and cost are, therefore, essential criteria of energy security.

The cost of acquiring energy should not limit present and future functions. Cost, however, is not only a matter of quantity. It is also related to norms in which the energy systems function. For instance, "affordable price" nowadays makes more sense if it is derived from market forces at levels with manageable consequences on other socio-economic activities and geopolitical factors (Bilgin 2010: 4–22).

> The interpretation of reasonable prices is somewhat less clear as it changes over time and is perceived differently by energy producers and consumers (Bielecki 2002: 237).

This is why political forces from demand and supply sides try to affect prices through changes in supply and demand, regulations (e.g. environmental incentives and taxation), financial means (e.g. future funds), technological breakthrough, reserve keeping and so on; while trying to benefit from external influences related to weather conditions, political instability, economic fluctuations, strikes and terrorist attacks.

Energy security concerns a variety of actors from individuals to states and societies. It appears, first and foremost, as a matter of nation states. Concerns on energy security provoke conflicts and promote cooperation both of which stem from priorities of nation states. The link of energy security to institutions is also significant. This is highly related to development of markets. Which one is more effective regarding energy security: States or institutions?

> The debate on energy security can be categorized in terms of two story lines; that of markets and institutions claiming that the rising influence of international markets is the key change since 1970s and that of regions and empires (Youngs 2009: 6).

The answer, therefore, lies in the characteristics of the concerned time interval.

What About the Twenty-first Century?

Neither of the story lines completely fit the twenty-first century, in which states and institutions confront regional, national, inter-national and trans-national challenges along with many other non-state actors.

> While in the past, national energy agendas have focused on supply security which has been achieved using domestic resources, today environmental and market liberalization debates increasingly dominate policy and influence the fuel mix in new directions (Chalvatzis and Hooper 2009: 2704).

This is why there is a need for a new perspective which combines the premises of these storylines for at least energy security in the twenty-first century derives its essence from regions and nation-states as much as from institutions and trans-national issues.

Having defined the evolution of energy security, it is now necessary to elucidate the term of energy transitions.

> The concept of energy transitions is based on the notion that a single energy source, or group of related sources, dominated the market during a particular period or era, eventually to be challenged and replaced by another major source or sources (Melosi 2006: 3).

Shifts to coal in the eighteenth and the nineteenth centuries, and to oil in late nineteenth and early twentieth centuries, led to significant changes in global political economy. The story lines were those of regions, empires, firms and nation-states simply because of the spatiality of coal and oil. The twenty-first century indicates a close relationship between international energy markets, regional and global state and non-state actors some of which gain strength from the oil economies of the last century, where as the others appear as rising powers owing their positions to recent and forthcoming dynamics. It is therefore possible to find a link between changing characteristics of energy security and energy transitions on the one hand; the rise and fall of regional and global powers on the other.

By moving on from this assumption; my aim in this study is to locate the characteristics of energy security within the notion of energy transitions, to identify idiosyncratic features of the twenty-first century, and to conclude whether current energy transition has been leading to drastic changes in regional and global relations. I intend to build upon my previous conceptualization of the New Energy Order (NEO) that was based on four general premises:

1. shift to diversified energy mix;
2. increase of actors in number and type;
3. ncrease of challenges and dilemmas in number and type;
4. rise of new global (such as Russia and China) and regional (such as Iran and Venezuela) power incumbents (Bilgin 2010: 4–22).

The New Energy Order necessitated feasibility, accessibility, sustainability and transparency (FAST) as the principles which defined energy security from a broader perspective. The FAST principles were thought to compromise the right of actors to develop their energy mix with the restraints to be elaborated in this chapter (Bilgin 2010: 4–22). Within this perspective; *Feasibility* defined the right of every actor to benefit from a variety of resources so far as the cost and externalities of the chosen fuel type do not limit their own economic activities. *Accessibility* guaranteed treatment of all energy products (from fossil fuels, to nuclear, renewables, hydrogen and others) equally and fairly with regard to the capability of actors to develop their own energy mix. *Sustainability* assured the right of every actor to benefit from a variety of resources so far as their choice did not endanger the capacities of other actors and future generations. *Transparency*, supported equitable and efficient use of energy resources, and necessitated full commitment of actors to become open to peer auditing and cross-checking to avoid free riders.

In this study, I deepen my analysis on each of these characteristics with regard to the link between energy security and energy transitions on the one hand; regional and global relations on the other. I focus on similarities and differences of energy transitions from a historical systemic perspective. The dominant energy type in particular eras can be considered as one of the most, if not the most, significant reasons which have shaped the concept of energy security. Shifts from one energy type to another results in drastic systemic changes in global political economy and vice versa. Understanding the shifts from one dominant energy type to other, is a key for elucidating the premises of energy security in particular time scales. This is how I analyze energy transitions to conclude on the characteristics of the twenty-first century in terms of the relations between energy and international security.

The second part after this introduction is entitled "Transitions to Fossil Fuels." It focuses on transitions to coal and oil, and addresses the role played by energy in colonial, regional, international and global relations from the eighteenth to the twentieth centuries. The third part identifies the current shifts under the title "Transition to the New Energy Order in the twenty-first century." It refers to contemporary challenges and points to continuities, discontinuities and changes in the twenty-first century. This analysis is meant to make more sense of the concept of the New Energy Order. This section first focuses on the shift to a diversified energy mix. It indicates the quantitative and qualitative factors which result in more use of nuclear and renewables along with sustained significance of fossil fuels. It then addresses the increase of actors in number and type such as the rise of the new seven sisters and selected countries. This part finally focuses on the increase of challenges and dilemmas such as climatic change and conflicts arising from commitments to curb carbon emissions; ecological and industrial consequences of 2010 BP oil spill; the relation between bio-fuels and food security; and the difficult encounter of nuclear energy with the concerns on nuclear weapon proliferation. The conclusion matches the data, facts and selected incidents with the premises of the New Energy Order and then it looks into the FAST principles with regard to priorities and restraints of selected actors. Finally, the conclusion introduces a FAST matrix, and maps the energy challenges and security concerns of the actors within the tracks of feasibility, accessibility, sustainability and transparency in order to locate fields of conflict and cooperation on selected problems.

Transitions to Fossil Fuels

Transition to Coal

The ways how we extract, use and consume energy generate idiosyncratic political and economic features differentiating one era from the other. This has been true since the very beginning of humankind. The discovery of making fire by primitive human beings shifted the balance of nature from systemic cycles of living organisms to energy based impact. The supply side of energy sustained its significance since then. Conversion of heat into electricity did not only accelerate scientific advances, but also raised the impetus of demand side on energy security. "This created a consumer society which considered electricity and other forms of energy as commodities that shall be available upon demand" (Andrews and Jelley 2007: 1–2). It is in fact possible to talk about demand and supply side aspects of energy which prove to be effective in the course of history. Both quantitative (changes in energy consumption) and qualitative (how the selected energy types influence and are influenced by the technical, economic, political, environmental and social forces) factors interact with energy transitions (Andrews and Jelley 2007: 1–2).

Historically, transition from one energy type to another caused significant quantitative shifts reflecting the changing modes of living. Cook (1971: 135), within this regard, categorized 6 terms of human history (primitive; hunting; primitive agricultural; advanced agricultural; industrial; technological) on the basis of daily energy consumption. In primitive times, when humans were only concerned with survival, with no systematic ability of hunting and cultivating, and with no control over fire, daily energy need was 2000 kcal per capita for nutrition. In the hunting period, characterized by the control of fire, ability for systematic hunting and the lack of agriculture, daily energy need per capita increased to 5,000 kcal for heating and cooking. In the primitive agricultural era, based on crops with support from animals, daily energy need per capita reached 12,000 kcal.

Advanced agricultural era was characterized by mechanization of agricultural techniques which allowed mass production with surplus. Daily energy need per capita in this era increased up to 26,000 kcal. The boost in agricultural production allowed more trade with other goods. Energy consumption tripled from advanced agricultural to industrial era which was about 77,000 kcal in average until the end of the nineteenth century. The twentieth and early twenty-first centuries were characterized by the technological period in which energy consumption per capita was about 230,000 kcal in advanced societies (Cook 1971: 135).

Energy transitions are significant not only because of the quantitative changes they stem from, but also due to the qualitative consequences they cause. Demand and supply side features influenced, and have been influenced by, the characteristics of societal and international relations in due course.

As to carbon fuels; the industrial revolution from the eighteenth to nineteenth centuries resulted in extensive use of coal by leading to the transition from labor based economy to machine based textile and iron industries. Changes in production allowed the rise of the United Kingdom, then continental Europe and North America as economic powers.

> Once nations had secured access to coal, the aim was to control large international (colonial) markets ... which served as outlets for products delivered by efficient coal-fueled domestic industries (Weissenbacher 2009: xix).

Coal fuelled steam power enabled a boost in production capacities and developed means of transportation. Extensive use of coal caused environmental degradation. Rapidly industrializing England of the nineteenth century, for instance, suffered from environmental and health problems which were caused by the clouds of smoke released from the increasing use of coal.

> The real solution to this was not found until after the Second World War, when coal was banned for domestic use in London, and this did not occur until petroleum products became available and began replacing coal (Adams 1982: 12).

Coal, in the mean time, helped industrializing countries to dominate world resources and markets, acquire wealth, and then transform the gains into technological breakthrough based on oil. Western powers gained a world wide supremacy by the competitive and military advantages of their coal fueled steam power against the wind powered ocean vessels. India became a colony; China remained under control; Africa became partitioned among European coal countries; the UK secured its influence in Australia and New Zealand (Adams 1982: 301–46). The USA, in the mean time, engaged coal fueled expansion in Latin America and North Pacific by dominating Hawaii and Philippines on the one hand, and by forcing Japan to allow construction of coal stations and access to Japanese markets, on the other (Adams 1982: 301–46).

Transition to Oil

Coal based steam power ships and trains in the late nineteenth century accelerated the economic progress and led to internal combustion engines and electrical power generation systems both fueled by oil. A Russian engineer, F.N. Semyenov, drilled the very first oil well in Azerbaijan in 1848 (Holland and Provenzano 2007: 25). The first method of drilling, (cable tool drilling was made by a rope connected to a wood beam has a drill bit attached to the end) for oil in the modern era was introduced by Edwin Drake in the US in 1850s (Fanchi 2004: 177).

> Developed in France in the 1860s, rotary drilling was first used in the US in the 1880s because it could drill into the soft rock formations of the Corsicana oil field in Texas.

The use of rotary drilling as "gusher" increased average oil production of a well from a few dozens to 100,000 in 1901 at Spindletop, near Beaumont Texas (Holland and Provenzano 2007: 25). Rotary drilling therefore helped the boost of oil production in the US which became the world's only supplier of crude and refined oil. The Standard Oil Company, which was founded by John D. Rockefeller as a business trust in 1882, emerged as an American giant company. In the mean time, Russian production in Azerbaijan, mainly from Baku fields, increased rapidly by the technological support from Ludwig and Robert Nobel; the brothers of Alfred Nobel. Under Nobel family's leadership, tsarist Russia became the world's second oil producer by challenging Standard Trust's overwhelming grip over international markets (Maugeri 2006: 11). The French branch of Rothschild family contributed to rise of Russia as an oil power when they constructed a railroad from Baku to Batumi (Black Sea port which nowadays takes place in Georgia) to transport kerosene in 1883. Rothschild family furthermore obtained production concessions and built refineries and became the second biggest oil company in the region after Nobels (Maugeri 2006: 11–12). While Standard Oil, Nobels and Rothschilds competed to dominate Asian markets, it was Marcus Samuel, from England, who established sound commercial networks in Asia and built advanced oil tankers capable of passing through Suez channel. This led to the rise of Samuel's Shell Company. Royal Dutch Company's discovery of oil reserves in East Sumatra (which nowadays takes place in Indonesia) made the company not only a competitor to big oil companies, but also a potential partner for Shell who had an advanced infrastructure of refineries, storage facilities, pipelines and vessels, yet lacked enough oil production agreements to comply with this huge capacity. The two companies indeed merged in 1907 and created Royal Dutch Shell by the efforts of Calouste Sarkis Gulbenkian who was an Istanbul born Armenian oil entrepreneur, with British citizenship (Maugeri 2006: 11–12).

Starting from 1899, Marcus Samuel (president of Shell at the time) had tried to convince British Navy to switch to oil from coal. According to some perspectives, what Samuel had asked was related to his concern with oil to make profits because this shift would create a new market. For some others, transition from coal to oil based navy would mean a strategic military advantage. In fact both approaches would prove valid. It was, however, the military aspect which had an impact on the course of political history. Winston Churchill, who was deeply concerned with the rising power of Germany's coal based navy, founded a commission to negotiate possible outcomes of Britain's decision to switch to oil based navy. The British navy converted from coal to fuel oil in 1910 and by following the precondition of securing its own oil resources as set by Churchill. In 1914, Britain acquired 51 percent of Anglo Persian Oil Company (APOC) which was originally formed in 1908. Middle East would shortly become a main concern for other countries which were in search of oil supplies. In cooperation with Calouste Sarkis Gulbenkian, Ottoman State established Turkish Petroleum Company (TPC) in 1910. Gulbenkian (40 percent), Ottoman State (35 percent) and Deutsche Bank (25 percent) were the shareholders. Shell obtained 25 percent from Gulbenkian whose share decreased to 15 percent after the transfer. TPC became Iraq Petroleum Company in 1914. Gulbenkian became the most predominant figure in Middle East's political history (Blair 1976: 31–4). TOC was restructured from 1913 to 1914 by re-allocation of quotas among the APOC (currently BP), the Royal Dutch Shell, and the German interests. Gulbenkian transferred shares to companies and he decreased his stake from 15 percent to 5 percent, yet kept a leading role to conclude the negotiations (Blair 1976: 31–4). After the break up of Ottoman Empire following the First World War, Gulbenkian led the "Red Line Agreement" of 1928. The agreement marked the former territories of the Ottoman State to include oil rich regions. It made the signatories (APOC, Royal Dutch Shell, Compagnie Française des Pétroles and the Near East Development Corporation composed of American companies) seek concessions only as part of this group within the red line and ensured 5 percent of share holding for Gulbenkian (CBF 2010). Consequently, TPOC was turned into Iraq Petroleum Company. This agreement did not only secure the concessions of big oil companies but also drastically shaped the political history in the Middle East.

Economically, oil boosted its economic significance after Henry Ford introduced Model T of first automobiles in 1908. "During the years prior to the First World War, oil was becoming an integral part of national economies without anyone taking notice, but during the Second World War, when success in combat depended on a steady and reliable flow of oil to fuel military vehicles, tanks and fighter planes, it was noticed" (Nersesian 2007: 132). Second World War would definitely become characterized by strategic significance of oil. Just like any other state, Hitler's Germany's main goal was to secure oil. Germany developed a technology to produce gasoline from coal, invaded Romania, entered North Africa and moved troops to seize Baku to control oil fields. Japan attacked Pearl Harbor because of the oil and steel embargo of the US. Germany lost the war partially because it failed at securing oil supplies. Japan collapsed extensively because of being deprived of the means to balance the development of nuclear technology in the US which would make the country surrender after Hiroshima and Nagasaki atomic bombings respectively on 6 and 9 August 1945.

> The international oil industry expanded quickly after the Second World War, largely from the base of low-cost oil reserves discovered in Venezuela and a handful of countries in the Middle East, by a handful of US and European countries, later dubbed 'The Seven Sisters' by Enrico Mattei, head of Italy's national oil company AGIP (Parra 2005: 5).

These companies were Shell; British Petroleum (formerly APOC); Exxon, Mobil and Chevron (successors of the Standard Oil); and Gulf and Texaco (companies which were originally founded to keep Standard Oil away from Texas). The Seven sisters dominated international oil politics until 1973 oil crisis.

Oil producing countries, in the mean time, became effective in regional politics by virtue of revenues they extracted from oil business.

> The most notable example of using oil as a political weapon to satisfy national interests occurred during 1960s with the formation of OPEC. Five major oil producing countries (Iran, Iraq, Kuwait, Saudi Arabia and Venezuela) primarily in the Middle East decided to take direct control of their oil industry and pricing away from the major oil companies … Each member agreed to own at least 50 percent of all oil operations in its respective country. Furthermore oil producing countries have provided assistance by supplying liberation movements with military support. For example during the 1970s members of the Arab Summit Conference assumed a leadership role in anti-colonial and antiapartheid struggles in Zimbabwe and South Africa by providing financial assistance to the liberation movements. Embargos have been used by oil producing countries, and this political tactic has also been used against them. Throughout history, a numerous organizations and political strategists have called on governments to use oil to seek political change. Within the past ten years, calls have been made for an international embargo, on Iraq and Nigeria, in order to place pressure on the military regimes (Falola and Genova 2005: 20).

Oil politics combined with the fragile structure of the Middle East, which in fact stemmed from the oil concerns of big powers, pushed the Middle East as the center of geo-political maneuvers. "Concerns over energy security have probably reached their peak during the 1970s when the world economy struggled to overcome the damaging effects of the oil crises of 1973–74 and 1979–80. These were triggered by inflation, which ultimately led to economic recessions involving substantial losses of Gross Domestic Product (GDP) and high unemployment" (Bielecki 2002: 236).

The twentieth century was characterized by the impact of oil on international political economy as shown by Table 2.1 (Morse et al. 2001: 97).

Table 2.1 The Oil Impact in the Twentieth Century

Oil related security flaws	First oil crisis (October 1973)	Second oil crisis (December 1978–October 1980)	Iraq–Iran War	Gulf Crisis (August 1990)	Market Conditions 2000
Cause	Fourth Middle East war embargo by Arab oil producers	Iranian revolution, rapid oil production decreases in Iran–Iraq War	Iraq attacks Iran	Iraq invades Kuwait	1999 OPEC Agreement and Low Investment
Supply decrease period	6 months	4 months	5 months	7 months	12 months plus
Supply decrease magnitude	4.3–4.5 million B/D (2 months) 2.2–2.6 million B/D (2 months)	5.3–5.6 million B/D (2 months) 3.8 million B/D (2 months)	3.7–4.1 million B/D (2 months) 2.5–3.0 million B/D (3 months)	5.0–5.3 million B/D (2 months) 4.0–4.7 million B/D (3 months) Total loss approx 400–500 million barrels	Over 1 billion barrel sustained OPEC production cuts
Excess production capabilities	About 3.75 million B/D	About 4.55 million B/D	About 6.70 million B/D	About 6.20 million B/D	1.0–2.0 million B/D
Number of days of petroleum stocks in OECD	Public: 0 Private: 70 days	Public: 7 days Private: 65 days	Public: 9 days Private: 77 days	Public: 25 days Private: 61 days	Public: 28 days Private: 53 days
Petroleum market structure	Majors posting price system. Majors' rights in long-term crude contracts.	Sales pricing system by governments of oil-producing countries. Long-term contracts with oil-producing countries	Sales pricing system by governments of oil-producing countries. Long-term contracts with oil-producing countries.	Market-linked pricing system. Development of oil futures market. Term contracts with oil-producing countries and expansion of spot transactions.	Market linked pricing system. Active oil futures market. Term contracts tied to spot transactions.

Note: In the Gulf Crisis, reduced crude oil supplies continued even after the war had ended until Kuwaiti production recovered.
Source: Morse et al. 2001: 97.

Table 2.2 Crude Oil Prices (US Average, $/bbl)

Year	Nominal	Inflation Adjusted	Year	Nominal	Inflation Adjusted
1970	3.39	19.04	2000	27.39	34.65
1973	4.75	23.13	2001	23.00	28.32
1974	9.35	41.27	2002	22.81	27.62
1978	14.95	49.99	2003	27.69	32.82
1979	25.10	74.67	2004	37.66	43.42
1980	37.42	99.11	2005	50.04	55.80
1990	23.19	38.57	2006	58.30	63.02
1995	16.75	23.96	2007	64.20	67.37
1997	18.64	25.32	2008	91.48	92.31
1998	11.91	15.93	2009	53.48	54.24
1999	16.56	21.62	2010*	70.67	70.84

Note: * Average as of July 2010.
Source: IDC 2010.

In October 1973, OPEC's decision to impose an oil embargo to challenge the decision of the US to grant emergency aid to Israel during the Yom Kippur War resulted in the first oil crisis with hiking prices (OPEC 2009).[1] Nominal oil prices remained at an average of $12–$13 from 1974 to 1978 as can be followed from the Table 2.2 (IDC 2010). The second oil crisis occurred in 1979 on the very eve of the Iranian revolution in which demonstrations against Shah Reza Pahlavi caused a decline in Iran's oil production. Iraq's invasion of Iran in 1980 halted oil production in Iran. Oil prices increased from $14 in 1978 to $37 in 1980. Higher oil prices promoted exploration and production in Mexico, Nigeria and Venezuela (Falola and Genova 2005: 179–236). Increase in production resulted in a low price period until 1986. These years of decline made the USSR suffer from decreasing revenues in an era characterized by increasing military and space expenditures.

Russian Federation could not manage at increasing production from 1990 to 1996 following the disintegration of the USSR. This failure contributed to price recovery which became effective since Iraq's invasion of Kuwait in 1990. Prices were negatively affected by 1997 and 1998 Asian financial crises and remained almost stable at about $16 in 1999. Relatively high average oil prices and increasing demand in the first decade of the twenty-first century made energy exporter countries more advantageous when compared to previous epochs. Energy in the twenty-first century, therefore, deserves a further scrutiny, with regard to power, wealth and security concerns of various actors.

Transition to the New Energy Order in the Twenty-first Century

The twenty-first century came along with a myriad of issues and challenges with which nation states could barely cope on their own. Energy security nowadays combines conventional criteria (sufficient amount of source at correct place and time and with affordable cost) with diverse issues of environmental and socio-economic factors (Bilgin 2010: 4–22). For instance, the emergence of anthropogenic climate change as an increasingly important energy policy made governments broaden their energy security agenda to minimize risk-prone fuels or suppliers while in the case of climate change mitigation to reduce greenhouse gas emissions from the burning of fossil fuels (Lefèvre 2010: 1636).

> A long-term energy policy would include a broad combination of measures and incentives that would stimulate domestic production, energy efficiency and conservation, promote environmentally-friendly technologies and eliminate any political barriers to global energy markets (Salameh 2003: 136).

Multitude of state and non-state actors, which are trying to take advantage of this shift in order to consolidate wealth, power and security, makes this picture more complex.

Energy security in the twenty-first century includes specific types of conflicts, antagonisms and dilemmas. Problems arise from a global transition to a new energy mix in which renewable and nuclear energy increase their significance along with rising shares of fossil fuels. The negative effects of bio-fuels on food security; domestic confrontation between hydro-energy and local ecological systems; limited contributions that might be obtained from wind, sun and hydrogen; and a myriad of unresolved issues raise additional questions on outcomes of this new energy order in which countries are obliged to secure their energy needs while coping with economic growth, environmental restraints, socio-economic problems and political pressures (Bilgin 2010: 4–22). Amongst these, four issues deserve a special attention in terms of the new energy order:

1. Characteristics of the shift to diversified energy mix;
2. Carbon emissions and environment;
3. Nuclear security;
4. Food security.

Shift to Diversified Energy Mix

Energy security nowadays combines conventional criteria of amount, price, location and time with environmental and socio-economic restraints which together lead to a broader understanding.

What is the link between energy transition to diversified energy mix and energy security from this broader perspective?

> Considering the case of fossil fuels, the dominating energy resource over the course of human history, each successive transition from one source to another – from wood to coal, from coal to oil – has entailed a shift to fuels that were not

> only harnessed and transported more economically, but also had a lower carbon content and higher hydrogen content (Asif and Muneer 2007: 1394).

This can be considered as evolutionary energy transition. Under normal conditions, an evolutionary energy transition would transform the shifts from coal to oil, from oil to natural gas and from natural gas to hydrogen – or cold fusion – for instance. The world, however, entered the twenty-first century without a successful transition to a lower carbon – higher hydrogen content fuel as previously expected. Energy systems, with the exception of minor applications, are currently deprived of a lower carbon – higher hydrogen content fuel with desired qualifications of cost, amount, time and place. It is not possible to talk about affordable costs; nor has it become possible to store and transport these energy carriers easily and with few risks.

It is therefore possible to conclude that science and markets, so far, have failed at continuing the energy transition from a "higher carbon and lower hydrogen" content to a "lower carbon and higher hydrogen" content fuel as they had once done. This trend implies of a new era in which transition to hydrogen (or another energy carrier or resource) has been delayed for technological and economic reasons.

> In past decades, hydrogen advocates believed that a global 'hydrogen economy' would begin to take shape near the end of the twentieth century, and that pure hydrogen would be the universal energy carrier by the middle of the twenty-first century. Hydrogen may not completely attain that lofty status in that time frame, but it is certain to play a much larger role – directly as a fuel for fuel cells, indirectly as an increasingly large component of carbon based fuels such as methanol and other conventional fuels – in the decades ahead (Hoffmann 2002: 10).

This is yet an important failure, or a technological delay, because while this transition has remained out of practice, growing economies and increasing population took place with a boost in transportation and consumption. The expected increase in energy consumption and rising significance of environment points to a question whether it may be possible to respond to challenges by attributing more shares to renewable energy. Unfortunately, all available renewable energy resources (river hydropower, wind power, tidal power, solar power, power of the thermohaline circulation, etc.) in total can supply no more than 10 percent of global consumption if no groundbreaking technological development occurs, and which is not expected, in the next decades (Anastassia et al. 2008: 281–8).

If it is not possible to talk about an extensive transition to a lower carbon and higher hydrogen content fuel; and if renewables are not adequate to cope with the growing energy demand, then what will be the next dominant energy type? State and non state actors will be concerned with transition to a new energy mix in which environmental and geopolitical restraints are minimized according to case based

Table 2.3 World Marketed Energy Use by Fuel Type, 1990–2035 (quadrillion Btu)

Year	Liquids	Natural Gas	Coal	Nuclear	Renewables
1990	136,40	75,40	89,20	20,40	33,90
2000	155,50	91,00	92,30	25,70	41,60
2005	170,40	106,30	122,30	27,50	46,20
2010	172,90	117,30	130,26	27,64	52,05
2015	179,33	129,07	139,08	32,19	63,81
2020	186,05	141,25	152,40	37,43	73,37
2025	197,23	150,21	167,76	41,09	82,43
2030	210,03	155,77	185,65	43,86	91,23
2035	223,57	162,03	206,26	47,08	99,78

Note: Table illustrates the reference case and ignores extreme contingencies such as durable economic recession or boom.
Source: EIA, International Energy Outlook 2010: 11.

criteria. Oil will sustain its significance or at least it will be very difficult, and may be expensive depending on the demand, to switch fully to electric fueled or hybrid cars. However, significant shifts can be expected so far as oil prices remain at high levels. Growing energy demand, oil prices and environmental restraints generate a shift to a variety of resources. Table 2.3, within this context, demonstrates that, while the total energy demand is growing by 1.4 percent per year, the consumption of liquid fuels will remain at an average annual rate of 0.9 percent which will decrease the share of liquids in total energy from actual percentage of 35 to 30 percent by 2035 (EIA 2010: 11).

In contrast to liquids, renewables appear as the fastest-growing source with an average annual growth of 2.6 percent. The increase in the consumption of natural gas, nuclear and coal are also remarkable. The total consumption of nuclear energy will reach 47.08 quadrillion Btu (QBTU) in 2035 by moving from 27.64 QBTU in 2010. Natural gas consumption is expected to gain momentum until 2020 (with a gas demand increase of 1.8 percent per year), and then decelerate to 0.9 percent per year because of the increase in renewable energy sources, nuclear power, and, coal; if lignite does not confront extensive environmental regulations (EIA 2010: 11). Deceleration in the growth of natural gas will arise from the expected increase in production costs. This will encourage more use of coal which will gain a momentum after 2020. Coal consumption is expected to increase at an average growth rate of 1.6 percent per year from 2007 to 2035 (EIA 2010: 11).

Many projections on energy mix indicate the continuing significance of fossil fuels for the first half of the twenty-first century, yet with different shares in different countries.

Table 2.4 World Marketed Energy Consumption by Country Grouping, 2007–2035

Region/2007–2035 (quadrillion Btu)	2007	2015	2020	2025	2030	2035	Average change (%)
OECD	245.7	246.0	254.2	263.2	271.4	280.7	0.5
North America	123.7	124.3	129.4	134.9	140.2	146.3	0.6
Europe	82.3	82.0	83.0	85.0	86.5	88.2	0.2
Asia	39.7	39.7	41.8	43.3	44.8	46.3	0.5
Non-OECD	249.5	297.5	336.3	375.5	415.2	458.0	2.2
Europe and Eurasia	51.5	52.4	54.2	56.2	57.8	60.2	0.6
Asia	127.1	159.3	187.8	217.0	246.9	277.3	2.8
Middle East	25.1	32.9	36.5	39.1	41.8	45.7	2.2
Africa	17.8	20.8	22.5	24.6	26.5	29.0	1.8
Central and South America	28.0	32.1	35.5	38.7	42.2	45.7	1.8
Total World	495.2	543.5	590.5	638.7	686.5	738.7	1.4

Source: EIA, International Energy Outlook 2010: 9.

> In particular, coal is projected to remain the mainstay of electricity generation in many major economies [e.g. China and India] – especially where coal is the main indigenous and economically viable, source of energy (Garg and Shukla 2009: 1032).

Table 2.4, supports this argument and indicates a radical divergence between OECD and non-OECD countries in terms of the change in their energy consumption from 2007 to 2035.[2]

The year 2007 was a historical moment when the total energy consumption in non-OECD countries exceeded OECD countries by 1.5 percent for the first time. Table 2.4 shows that non-OECD countries will sustain this trend in the future with an average annual growth rate of 2.2 percent in contrast to OECD countries whose energy consumption growth will not exceed 0.5 percent per year from 2007 to 2035.

Table 2.4, which is based on IEO2010 reference case, suggests that the world energy consumption will increase 49 percent from 495 quadrillion Btu in 2007 to 739 quadrillion Btu in 2035 (EIA 2010: 9). The negative impact of the global economic recession has been considered as a reason of a decline of 1.2 percent in 2008 and 2.2 percent in 2009. It is however assumed that by 2015, most of the countries will be able to sustain their expected long-term growth rates before the recession. Accordingly, the world GDP is expected to grow 3.2 percent per year from 2007 to 2035. The difference between OECD and non-OECD countries is tended to grow with non-OECD economies averaging 4.4 percent per year and OECD economies 2.0 percent per year (EIA 2010: 9).

Non-OECD countries will take most of the advantage of this economic growth. Jim O'Neill, from Goldman Sachs, was aware of this trend and urged how important it was to have Brazil, Russia, India and China (BRICs) to act well adjusted to global system for the well-being of the World economy (O'Neill 2001). Actually, Russia is an accession candidate country to OECD; where as Brazil, China, India, Indonesia and South Africa are countries with enhanced engagement (OECD 2010). Regardless of their full membership to OECD, the BRIC terminology is useful for understanding the significance of these countries for global economic relations. This term, however, falls short of indicating the geo-political, geo-economic and security dimensions from an energy perspective. Economic growth would make less sense without energy security, which in fact is an essential element to become a global or a regional power. This is why there is a further need to map some countries, in addition to BRICs, as emerging global and regional powers in terms of their ability to benefit from the energy transition in the twenty-first century.

The New Seven Sisters, Global Actors and Regional Powers

The concept of BRIC countries is derived from an economic perspective. Defining the power incumbents of the twenty-first century would be more comprehensive from the energy security outlook for at least the growth of BRIC stimulates a rapid increase in energy demand, which in turn makes energy rich countries, and their companies, more advantageous when compared to previous epochs.

The increase in demand, which was highly related to China's and India's economic growth, triggered the hike of oil prices in the first decade of the twenty-first century.

> As the richest and fastest-expanding nations outgrow their energy supplies, and become more dependent on energy exporting nations that are often economically poorer and politically less stable, energy security has assumed a central role in global security and international relations (Nolan et al. 2004: 150).

This had occurred previously when prices increased because of 1973 and 1979 oil crises. What fundamentally differed in the first decade of the twenty-first century was the ability of many power incumbents (such as Russia, Iran and Venezuela) to transform oil and gas revenues into policy tools to consolidate state power along with wealth. This is a radical shift.

The so called seven sisters, which nowadays turned into the giant fours composed of ExxonMobil, Chevron (Texaco), Royal Dutch Shell, and BP (Amoco-Arco), shaped and took advantage of the oil era in the twentieth century. They did not only dominate their home markets through vertical integration, but also worked together to control international supplies of oil by excluding competitors through the cooperative ventures they formed to explore, produce and refine foreign reserves without involving in a price competition (Spero and Hart 2009: 340–41). Energy actors of the twenty-first century fundamentally differ from the twentieth century. This shift supports national companies which combine wealth and power with state's priorities.

A significant proof to this argument comes from rankings of the energy companies as shown in Appendix 2.1 (see the end of this chapter) (PIW 2009).

The table in Appendix 2.1 compares six different operational areas in which companies are assigned rankings according to selected criteria that include liquids and gas in terms of output and reserves, product sales, distillation capacity in addition to financial and other measures of size comprising revenue, net income, total assets and number of employees (PIW 2009).

Most of the new giants come from the so called non-Western world. Economically; many of them are from non-OECD countries. Politically and militarily; most of them are too far away from the Euro-Atlantic alliance. From business perspective; many of them are state or state controlled firms. A useful categorization on some of these companies has already been introduced by Hoyos (2007):

> When an angry Enrico Mattei coined the phrase 'the seven sisters' to describe the Anglo-Saxon companies that controlled the Middle East's oil after the Second World War, the founder of Italy's modern energy industry could not have imagined the profound shift in power that would occur barely half a century later, which comprises the rise of new seven sisters: Saudi Aramco, Gazprom, CNPC, NIOC, PDVSA, Petrobras and Petronas.

The seven sisters, nowadays face the reality of *the new seven sisters* from Saudi Arabia (Saudi Aramco); Russia (Gazprom); China (Chinese National Petroleum Corporation, CNPC); Iran (National Iranian Oil Company, NIOC); Venezuela (Petróleos de Venezuela, S.A. PDVSA); Brazil (Petrobras); and Malaysia (Petroliam Nasional Berhad, Petronas) (Hoyos 2007).

> The new seven sisters control almost one-third of the world`s oil and gas production and more than one-third of its total oil and gas reserves in contrast to the old ones (which shrank to four in the industry consolidation of the 1990s) produce about 10 percent of the world`s oil and gas and hold just 3 percent of reserves (Hoyos 2007).

These national companies challenge the conventional actors, the rules of the energy game and help their home countries increase efficiency in global political economy. Table 2.5 illustrates how regional and global power incumbents, with explicit state ideologies, have managed at establishing leading energy companies as listed in PIW rankings.

It is important to underline that these seven companies come from countries which somehow proclaim a sort of state ideology to become regional or global powers. Saudi Aramco cannot be detached from the political understanding of the royal family which is argued to be effective in some Gulf countries. Aramco's oil fields and the significance of this company for the Royal family play an important role not only with regard to regional relations but also in

Table 2.5 New Seven Sisters, Energy and State

Oil Company	PIW Ranking (2008)	Country of Origin	State Ownership	Country's Main Energy Concern Other Than Oil	Trans-boundary Externality of Energy Agenda	State Ideology	Power Incumbent
Saudi Aramco	1	Saudi Arabia	100	Natural gas	Carbon emission	Sunni Monarchy	Regional
NIOC	2	Iran	100	Nuclear and natural gas	Carbon emission – nuclear security	Shia regionalism	Regional
PDVSA	4	Venezuela	100	Hydroelectric	Carbon emission – Ecology	Neo-Bolivarian regionalism	Regional
CNPC	5	China	100	Coal and Nuclear	Carbon emission – ecology – nuclear safety	Authoritarian State-capitalism	Global
Gazprom	14	Russia	50,023	Coal and Nuclear	Carbon emission – ecology – European gas supply security	Oligarchic State-capitalism	Global
Petrobras	15	Brazil	32,2	Hydroelectric and biomass (ethanol)	Carbon emission – ecology – food security	Assertive globalization	Global
Petronas	18	Malaysia	100	Natural gas and LNG	Carbon emission	Islamic capitalism	Regional

Source: Author.

terms of the tense relations between Saudi Shi'ite (making 5–6 percent of Saudi population) and Sunni especially given the fact that Saudi Wahhabis actively reject all veneration of a man, even the prophet (Cordesman 2003: 206). In Russia; Gazprom appears a very contentious company which, somehow, supports Kremlin's policies without endangering firm interests (Bilgin 2009: 4482–91). Although there are many factions in Gazprom and Kremlin, the use of Gazprom as a strategic tool by government is obvious according to many observers (Rosner 2006). Gazprom's corporate initiatives in Europe, Central Asia, the Middle East and Africa extensively fit Russian state's geopolitical priorities (Bilgin 2007: 6383–94). China attributes a special mission to CNPC for consolidating its energy security in addition to other national energy companies which go global (Kong 2010: 61). Investments of CNPC support foreign policy priorities of the Chinese state in Central Asia (e.g. Turkmenistan), Middle East (Iran) and Africa (e.g. Sudan). NIOC's encounter with Shia ideology and the Iranian state's political ambitions in and around the Gulf can be understood from an energy perspective (Stern 2007: 377–82). Iranian state would not be able to support a nuclear energy program, develop natural gas infrastructure, expand military capabilities and cope with domestic economic problems without NIOC. It is therefore possible to point out the indirect effect of NIOC on Iran's capability of expanding its power in Gulf especially in Southern Iraq. PDVSA enables Hugo Chavez to implement a sort of neo-Bolivarian politics for consolidating power in Americas. "The company, Chavez remarked, gave us a great opportunity to move ahead with the government's political and social agenda" (Kozloff and Chávez 2007: 34). He would not be able to adopt a radical discourse against the US and US interests without PDVSA. Brazilian state's share in the national energy company is the smallest when compared to others in the table. Brazil does not assert a challenging state ideology, but assumes a global actor's role by increasing efficiency within the existing system. Brazil, in search of becoming a permanent member in the UN Security Council, owes a lot to Petrobras' economic and technical (take under sea operations for instance) potential when it comes to acquiring the status of a global actor. Petrobras is one of the world's biggest oil producers but also the world's biggest bio-fuel producer (Slater 2010: 168). Malaysian government fully owns Petronas, which had once belonged to Royal Dutch Shell that was particularly concerned with extracting oil from British colonies. Petronas' current situation is much better than BP which has been seriously damaged by the financial consequences of the oil spill in Mexico Gulf as will be discussed later. Petronas, which provides roughly 45 percent of the government's revenue through taxes and dividends, has become even more important for the government which has been in search of increasing this contribution (Chew and Barta: 2010). Petronas proved to have a certain degree of independence from the government until recently. This may change on behalf of government which replaced the CEO of the company to assert more government control (Chew and Barta: 2010). Malaysia may, therefore, be taking the track of

Russia, Iran and Venezuela to benefit from the national energy company not only economically, but also politically and strategically.

Table 2.5, and the discussion built upon it, is not to assert that energy companies are promoting state ideologies. Rather it shows that if economic growth and energy security are two significant factors of becoming global and regional powers, then the state ideologies of these countries deserve further attention. It is not clear whether economic growth and energy trade may channel less antagonistic state ideologies, especially in case of China, Iran, Venezuela and Russia; or whether competition over resources may cause more regional and international conflicts. The clear fact stems from the rise of these seven and many other "non-Western" companies that challenge conventional actors and norms of global energy markets.

Environment: From Carbon Emissions to the BP Oil Spill

Environmental restraints are relevant for energy products with no exception since none of them are totally secure in terms of ecology, health and safety. It is however possible to indicate that commitment to these issues increases the energy production costs because they necessitate additional processes to minimize risks and negative consequences. This is true not only for fossil fuels, but also for alternatives such as wind, bio-fuels and nuclear. Countries' commitments to curb carbon emissions and an incident, 2010 BP oil spill, deserve a further attention within this regard.

It was in 1896 when Svante Arrhenius (a Nobel Prize winner Swedish scientist) had pointed out that doubling the carbon dioxide content in the atmosphere would cause a temperature increase of 4 to 6 Celsius degrees (Quaschning 2010: 30). In 1950s, it was scientifically proven that Carbon dioxide sequestration in the atmosphere was steadily increasing because of industrialization (Quaschning 2010: 30). Nowadays, Carbon dioxide concentration from fossil fuels is considered to be one of the main reasons which cause climate change (Bauen 2006: 893). "The concentration of gases that have been proven to cause global warming has increased considerably in recent decades by causing anthropogenic green house effect" (Quaschning 2010: 30). Energy policies, therefore, take global warming as a significant criterion of security. Restrictions on carbon emissions are likely to cause a negative impact on fossil fuels. Vice versa, less commitment to environment would mean more shares for fossil fuels.

The climate summit in Rio de Janeiro (3–14 June 1992) put emphasis on environmental restraints and sustainable development with Agenda 21 that included principles for attaining better environmental standards on an equitable basis (UN 1992). The Kyoto Protocol (11 December 1997) set binding targets for 37 industrialized countries and the European community for reducing greenhouse gas emissions which were supposed to amount to an average of 5 percent against 1990 levels over the five-year period 2008–2012 (UN 1992, 1998).[3] These measures, however, did not result in a positive impact as much as previously expected. "Most Kyoto signatories failed to reduce emissions [with partial exception of Central and Eastern Europe] and global Carbon emissions grew four times more quickly

Table 2.6 World Energy-related Carbon Dioxide Emissions, (billion metric tons)

2007–2035	OECD	Non-OECD	World Total
2007	13,7	16	29,7
2015	13	18,5	31,5
2020	13,1	20,7	33,8
2025	13,5	23	36,5
2030	13,8	25,5	39,3
2035	14,2	28,2	42,4

Source: International Energy Outlook 2010: 8.

between 2000 and 2007 than they did between 1990 and 1997" (Bickel and Lane 2010: 11). Copenhagen Climate Summit in December 2009 aimed to come up with a new protocol to succeed Kyoto Protocol after 2012. Commitments to curb carbon emissions remained even more limited when compared with Kyoto. "The accord, which falls short of a binding treaty sought by many nations, sets a goal of limiting global warming to below 2 degrees Celsius (3.6 Fahrenheit) above pre-industrial times, and leaves each nation to set its own targets for 2020" (Doyle and Wynn 2010). The Copenhagen accord promised $100 billion climate aid annually by 2020, and $30 billion from 2010–2012 to make poor nations more willing to commit themselves to curb their emissions (Doyle and Wynn 2010). This aid was considered to be unsatisfactory as well.

Global carbon dioxide emissions in 2009 fell 1.3 percent to 31.3 billion tons in the first year-on-year decline in this decade due to global economic crisis and rising investments in renewable energies (Eckert 2010). Accordingly, China in 2009 ranked the first in terms of carbon dioxide emissions with 7.43 billion tons (6.81 billion in 2008), followed by the US with 5.95 billion (6.37 billion 2008) (Eckert 2010). Russia, India and Japan came after the USA. As to future; carbon dioxide emissions which were reported to be 29.7 billion metric tons in 2007 are expected to increase to 33.8 billion metric tons in 2020 and 42.4 billion metric tons in 2035, with growing amount of emissions from non-OECD countries as can be seen from Table 2.6.

How can the country positions, *vis-à-vis* climate change, be identified? In general there are three groups of countries which come up with a consistent strategy to cope with climate:

1. The Western countries seem to be willing and readily able to adjust in response to rapid and serious climate change;
2. Some developing countries (mainly the BRIC group) are willing to adjust, but facing significant economic hardship without external assistance and protection;
3. Some other countries are unwilling and/or unable to play a part in combating climate change (Nuttall and Manz 2008: 1247–59).

Box 2.1 BP Oil Spill and Deep Sea Exploration

Another significant pressure, on fossil fuels, particularly on deep shore oil production, arose in the after math of the BP oil spill in Gulf of Mexico. Deep water oil exploitation had become feasible since the prices went beyond $65 per barrel. Ecological concerns became very effective, and turned out to be an impediment in front of deep oil production, specifically after the BP accident in Gulf of Mexico. Explosion of deepwater horizon on 20 April 2010 killed 11 people and caused contamination of sea water with devastative negative effects on ecological systems (Pidd 2010). This accident was 4–5 times bigger that the so called biggest Exxon Valdez disaster of 1989 causing 40–120 m liters (260–750 thousand barrels) of oil spill. If Exxon Valdez disaster, which had occurred when Exxon Valdez oil tanker hit a reef in Alaska on 24 March 1989, was the most devastating human-caused ecological disaster to be publicly known; then BP oil spill is going to be one of the worst of worst, if not the worst of worst, ever. Exxon Valdez accident made the oil industry to take important measures at tanker transportation. The BP oil spill in Gulf of Mexico is not only enormous when compared to Exxon Valdez, but also is more destructive for the oil industry both in financial and investment terms. It may even be concluded that BP oil spill in Gulf of Mexico hit the oil industry as a whole, discouraged massive investments in deep sea unless new technologies can secure prompt control of accidents.

Carbon emissions will continue to be a thorny issue for at least it seems to be very difficult to find an equitable way which satisfies different concerns of Western countries, BRIC, and other developing countries. The pressure on curbing Carbon emissions make countries attribute growing shares to nuclear and renewable energy, yet with new risks and restraints.

Nuclear Energy and International Security

A decrease in the share of nuclear energy in total energy consumption was expected at the eve of the twenty-first century. There was a widespread concern over the possibility of reactor accidents, the disposal of nuclear wastes, and nuclear weapon proliferation (Bodansky 2004: 21). Above all, the nuclear power cost was higher than alternative fuels in real terms of the year 2001. Most of these assumptions would continue to be effective so far as oil prices remained relatively low from 1980s to September 2003 with a NYMEX average of $25 per barrel. Significance of nuclear energy started to increase following the historical peak of oil prices. As discussed earlier, the rally in oil prices from 2003 reached a historical peak of $147 in July 2008, decreased $32 in December 2008 and then remained above $60 after October 2009.

In addition to relatively higher oil prices, environmental restraints, some of which have been addressed in the previous section, support the projection of IEA and EIA both of which presume more use of nuclear energy in the first half of the twenty-first century. Prices and environmental restraints are not the only reasons which promote more use of nuclear energy. Recent developments, along with mid-term expectations, are very effective. Technical improvements in safe reactor management, secure disposal of wastes, the possibility of using nuclear to produce hydrogen for fuel cell based transportation system, its practical use for desalinating sea water, its being a promising way of supplying steam and electricity for shale oil and oil sand deposits (e.g. those in Canada), the contingency of using thorium to increase affectivity of nuclear power plants at feasible terms, are among the many reasons which stem from the energy sector and raise the significance of nuclear energy (Hore-Lacy 2006: 38–127).

These characteristics support the shift to a global energy mix in which nuclear is likely to take more shares. As the table in Appendix 2.2 shows, nuclear power reactors will drastically increase and spread in the next decades (see the end of this chapter) (WNA 2010).

The dichotomy between the right of every nation to benefit from technology in equitable terms and the doubts on nuclear weapon proliferation turned out to be more serious once countries became more concerned with building new nuclear power plants. The safeguards and principles adopted by International Atomic Energy Agency (IAEA) and Nuclear Non Proliferation Treaty (NPT) impose standards and criteria to avoid military use of nuclear energy. It is yet a very contentious issue especially in the twenty-first century, when countries will be demanding more nuclear energy to respond to environmental restraints, relatively high oil prices, and their own energy security priorities.

Uranium enrichment, particularly separation and concentration of the fissile isotope U-235 and processing spent fuel for weapons usable plutonium are the main issues which raise doubts on nuclear energy programs of various countries including Iran, Libya and North Korea given the sale of sensitive technology and equipment to these countries by Pakistani scientist A.Q. Han; a transmission that became evident in 2004 (Holt 2009: 22). Since Pakistan remained out of the NPT, and Iran, who is a member to it, has been very keen to develop nuclear energy, these two countries became the nucleus of international concerns over the possibility of nuclear weapon proliferation.

Iran, nowadays, appears as a case which indicates the difficulty of comprising divergent interests between the right of a country to acquire nuclear energy and international doubts on the possibility of transforming nuclear energy into weapons of mass destruction. Iran actually accepts no intention to develop nuclear weapons and emphasizes its need to develop nuclear power plants to generate electricity and increase natural gas exports. Is it possible to justify Iran's interest from an energy perspective by referring to the premises discussed within this chapter? Energy demand in Iran is increasing. Iran, as the world's second richest natural gas country after Russia, occasionally imports gas mainly from Turkmenistan to satisfy its energy demand, and meet its OPEC oil quota by injecting natural gas in decaying oil wells. Nuclear power plants will definitely help Iran to boost natural gas production, recover the decay in mature oil fields and satisfy the domestic demand with no external dependence. However, Iran's political regime and tense relations with Israel and the USA make international community get concerned with Iran's nuclear agenda.

> At an international level, Iran has courted the particular antipathy of Washington by its sponsorship of international terror, its alleged interference in the affairs of its neighbors, Afghanistan and Iraq, and its failure to answer important questions about a nuclear program, that is supposedly intended to supply energy to a fast-growing population but which in practice may well continue to conceal an ambition to build a nuclear warhead (Howard 2004: 3).

The Iranian case, therefore, shows how international relations get more complicated when it comes to energy security, and especially to nuclear energy. Yet, it is not the only case as can be understood from Appendix 2.2 which shows that security issues on nuclear energy necessitate full transparency, governance and peer auditing on safety of reactors, disposal of nuclear wastes, and possibility of nuclear weapon proliferation.

Bio-fuels and Food Security

Bio-fuels help curbing carbon emissions while decreasing external dependence of countries. Just like any other alternative fuels, development of bio-fuels is highly related to oil prices and environmental restraints.

> The driving forces behind bio-energy development include not only its ability to compete with petroleum prices, but also its potential capacity to reduce global greenhouse gas emissions and to enhance farmers' income (B. Kumar et al. 2009: 265).

Their socio-economic contribution, however, goes far beyond that. Bio-fuels – if managed successfully – can help diversification of agricultural output, create employment and increase the scope of carbon credits (B. Kumar et al. 2009: 265–70). High oil prices and concerns over carbon emissions promoted cultivation of bio-fuel corps because some governments became more interested in subsidizing it through tax exemptions, consumption targets and technical aid. For instance global ethanol production tripled from 18 billion liters in 2000 to 46 billion liters in 2006, by concentrating in Brazil and the USA which together accounted 95 percent of production in 2007 (Msangi et al. 2010: 69–70). Bio-diesel production, which is relatively new when compared to ethanol, has increased five fold from 2000 to 2007 and reached 6 billion liters, 79 percent of which belonged to France and Germany (Msangi et al. 2010: 70). In addition to Brazil, the EU and the USA; Argentina, Indonesia, Malaysia and Thailand are the countries with minimum blending requirements or future targets for renewable fuels the production of which is expected to increase in three different scenarios as can be seen from Table 2.7 (Msangi et al. 2010: 68).

Table 2.7 Projected Demand for Feedstock Commodities for Bio-fuels in 2020

Crop	Region/Country	Baseline	Bio-fuel expansion	Drastic bio-fuel expansion
Cassava	ROW	660	6842	13684
Maize	EU	97	1086	2173
	ROW	2021	20511	41023
	USA	35000	130000	260000
Oil Seeds	Brazil	16	153	306
	EU	1563	14572	29144
	ROW	530	4211	8423
	USA	354	3017	6034
Sugar	Brazil	834	9014	18029
	ROW	163	1797	3595
	USA	265	3450	6900
Wheat	EU	1242	10703	21407
	ROW	205	2342	4685

Source: Msangi et al. 2010: 73.

The Baseline Scenario defines a situation in which demand is not extensively increased from 2010. Bio-fuel expansion scenario is obtained from the plans of individual countries to increase the share of bio-fuels to balance high oil prices. Drastic bio-fuel expansion assigns supremacy to fuels over food, which in turn negatively affects global food markets, food prices and malnutrition in developing countries. High oil prices are expected to lead to bio-fuel expansion. This may turn into drastic bio-fuel expansion if oil prices and environmental restraints drastically increase with less opportunity to benefit from other resources such as coal, gas, uranium and renewables.

Both of the scenarios result in significant challenges against food security. To start from current situation, the negative impact of large-scale ethanol production is important.

> One of the biggest concerns associated with the production of ethanol is related to its current reliance on starch from corn and wheat. This is perceived as a competition between fuel and food (Vermerris 2008: 31).

The impact of bio-fuels on food security has two aspects:

1. A direct affect occurs, once grain is converted from a product of food and feed to fuel;
2. An indirect affect occurs once the land used for food production is replaced by bio-energy corps (Vermerris 2008: 31).

In turn, food shortage and increasing food prices emerge as the combined effect.

Bio-fuels have a potential to contribute to energy security. However, the expected bio-fuel expansion may entail economic and ethical questions for at least the problems of nations suffering from extreme poverty; malnutrition and famine are likely to increase while the others can cover hiking food prices at the cost of an economic burden balanced with energy savings.

> While energy security concerns may well justify the development of bio-fuels, the proposed scale of development has raised questions about trade-offs between bio-fuels and food crops, as well as between bio-fuels and bio-electricity (Thomas et al. 2009: 1140).

Just like other challenges of the new energy order in the twenty-first century, bio-fuels entail trans-boundary problems with direct and indirect affects on third parties.

Conclusions

Energy transitions in the past three centuries fundamentally affected the characteristics of the world system. This study searched whether it might be possible to indicate the shifts in regional, international and global relations by identifying idiosyncratic characteristics of the transition in the twenty-first century.

Increasing energy demand mainly from non-OECD countries, relatively high oil prices and environmental restraints appear as the main factors which currently generate a further energy transition. Environmental restraints support continuation of the former transitions from "high carbon low hydrogen" content to "lower carbon higher hydrogen" content. Yet the increase in demand makes the actual transition deviate from this conventional course which was formerly characterized by a transition from wood to coal and from coal to oil, in conformity with the logic of moving on from "higher carbon lower hydrogen" content to "lower carbon higher hydrogen" content. Historically, and within this logic, natural gas, and then hydrogen could have become the dominant energy source (carrier in case of hydrogen) after oil. This did not happen until recently. A further diversification in the energy mix of particular countries is evident. This is to respond to environmental restraints and the increasing demand at affordable costs. Nuclear and renewable energy are likely to increase their share in total energy consumption. Fossil fuels will keep their significance so far as a technological breakthrough does not lead to an unprecedented abundance of a cheaper energy source. This shift however results in additional challenges such the link between nuclear energy and the doubts on nuclear weapon proliferation and the affect of bio-fuels on food security.

Energy security in the twenty-first century indicates a remarkable confrontation between the OECD and non-OECD countries. Non-OECD countries constitute the main drive for the economic growth, energy demand and carbon emissions. Energy rich non-OECD countries and their companies have consolidated wealth and power by benefiting from relatively high oil prices. The new seven sisters from Saudi Arabia (Saudi Aramco); Russia (Gazprom); China (Chinese National Petroleum Corporation, CNPC); Iran (National Iranian Oil Company, NIOC); Venezuela (Petróleos de Venezuela, S.A. PDVSA); Brazil (Petrobras); and Malaysia (Petroliam Nasional Berhad, Petronas) are the best examples of this rise. Private and state companies from OECD countries such as BP, Exxon Mobil, Shell, Statoil, Total and many others consider these companies as the competitors yet with new opportunities for cooperation. State and non-state actors, in the mean time, try to curb carbon emissions, respond to environmental restraints and prove a certain degree of respect to ecological systems. 2010 BP oil spill appeared as a good case which indicated how a massive failure on ecological management could create a huge financial burden with negative effects on deep sea drilling industry.

This study, therefore, pointed to some similarities (continuing significance of fossil fuels and spatiality of energy geopolitics) and some significant differences (new actors, norms, threats and risks) of the twenty-first century when compared to previous epochs. From this stand point, twenty-first century could be explained by the concept of the New Energy Order, based on four general premises:

1. shift to diversified energy mix;
2. increase of actors in number and type;
3. increase of challenges and dilemmas in number and type;
4. rise of new global (such as Russia and China) and regional (such as Iran and Venezuela) power incumbents (Bilgin 2010: 4–22).

These systemic and conceptual changes were matched with many facts and incidents stemming from either of these characteristics. The most significant ones, elaborated within this chapter, can be clustered as:

1. Environmental degradation and climate change which necessitated curbing carbon emissions;
2. Ecological disasters and accidents arising from the energy sector (e.g. Exxon Valdez oil spill in 1989 and BP oil spill in 2010) which necessitated sustainable management of energy production, transportation and consumption;
3. Concerns on the renaissance of nuclear energy (safety, waste management; nuclear weapon proliferation) as energy from uranium seems to be a very practical way of coping with rising oil prices while curbing carbon emissions;

Table 2.8 FAST Matrix: Selected Energy Challenges and International Security

Selected Challenge	Externality	Claim on *Feasibility*	Claim on *Accessibility*	Claim on *Sustainability*	Claim on *Transparency*	Main Area for improvement
Iran's Nuclear Energy Program	Possibility of nuclear weapon proliferation	Iran	Iran	Iran and International Community	International Community, the USA, the UN	Transparency
BP Oil Spill	Ecological disaster and doubts on deep sea oil drilling	BP	BP	BP, International Community, the USA	BP, International Community, the USA	Sustainability
Fossil fuels with high carbon emissions	Conflicts between developed and developing nations	non-OECD Countries	non-OECD Countries	International Community and OECD countries	International Community and OECD countries	Accessibility
Bio-fuel Expansion	Ecological impact, challenge against food security	Brazil, USA, Germany, France	Brazil, USA, Germany, France, Argentina, Indonesia, Malaysia and Thailand	International Community and Poor nations	International Community and Poor nations	Sustainability

Source: Author.

4. The link between bio-fuel and food security (too much of lands attributed to bio-fuel production limits foodstuff).

It is not easy to harmonize spatial and finite characteristics of energy and inequitable commitments arising from carbon emissions with the priorities of a myriad of actors seeking interest maximization. Rational choices of individual actors cause negative consequences and externalities which transcend borders. Table 2.8 maps these challenges within the FAST matrix, transform them into practical tracks, and builds a table of challenges and conflicts to locate the areas on which the actors can negotiate, improve conditions, or minimize risks and threats.

Accordingly, Iran's nuclear energy program raises doubts on the possibility of nuclear weapon proliferation. Iran claims the right to access a feasible energy mix which would serve for the sustainability of world resources and facilitate curbing carbon emissions. The main area of improvement appears as the issue of transparency from this perspective.

BP oil spill exemplifies an ecological disaster which leads to doubts on deep sea oil drilling with negative influence on supply. Deep sea drilling is accessible for BP who considers these operations feasible. BP has already justified exploration and production with a strict commitment to sustainability. BP oil spill in Gulf of Mexico turned these justifications upside and down. BP, assuming responsibility, is in need of recovering the ecological damage and proving its commitment to nature. This appears as an area of improvement within the track of sustainability.

Carbon emissions are considered to be a significant reason for global warming. There is a need for voluntary initiatives to curb emissions given the loose characteristics of post-Copenhagen era. Many non-OECD countries, mainly China, claim to use their domestic coal resources to decrease costs and consolidate their national energy security. Feasibility is the main justification of this approach, while sustainability is a concern of the international community. China can balance environmental effects of coal with a diversified energy mix so far as other types of energy from nuclear to natural gas are accessible. Accessibility appears as the most significant track within this context. Transparency is also important, yet seems to be an issue of an ongoing process to be evaluated in terms of China's energy mix.

Bio-fuel expansion contributes to energy security by diversifying the energy mix stemming from accessibility. Yet it challenges food security which makes the countries mentioned in the table stand at odds with the international community as much as with poor nations. Sustainability from this perspective appears as an area for improvement.

So to conclude; the twenty-first century differs from the previous ones in terms of the actors, challenges and opportunities. Many of the challenges, some of which are elaborated within the FAST matrix, necessitate trans-national cooperation among the concerned state and non-state actors. The New Energy Order, within this context can lead to equitable use of resources so far as the actors attribute a similar meaning and priority to feasibility, accessibility, sustainability and transparency.

Notes

1 Other members which joined OPEC's five founding members are: Qatar (1961); Indonesia (1962) – suspended its membership from January 2009; Socialist Peoples Libyan Arab Jamahiriya (1962); United Arab Emirates (1967); Algeria (1969); Nigeria (1971); Ecuador (1973) – suspended its membership from December 1992–October 2007; Angola (2007) and Gabon (1975–1994) (OPEC 2009).

2 OECD member countries, as of March 10, 2010, are: Australia, Austria, Belgium, Canada, Czech Republic, Denmark, Finland, France, Germany, Greece, Hungary, Iceland, Ireland, Italy, Japan, Luxembourg, Mexico, Netherlands, New Zealand, Norway, Poland, Portugal, Slovakia, South Korea, Spain, Sweden, Switzerland, Turkey, the United Kingdom and the USA. Although Chile is a member since 7 May 2010, it is not reflected in tables with references to IEO 2010.

3 The Kyoto protocol entered into force on 16 February 2005.

Appendix 1.1 Top 50 Energy Companies

2008 Rank	2007 Rank	PIW Index	Company	Country	State Ownership (%)*
1	1	29	Saudi Aramco	Saudi Arabia	100
2	2	33	NIOC	Iran	100
3	3	37	Exxon Mobil	US	
4	4	49	PDV	Venezuela	100
5	5	53	CNPC	China	100
6	6	55	BP	UK	
7	7	65	Royal Dutch Shell	UK/Netherlands	
8	8	87	ConocoPhillips	US	
9	9	90	Chevron	US	
9	10	90	Total	France	
11	11	92	Pemex	Mexico	100
12	14	101	KPC	Kuwait	100
13	12	103	Sonatrach	Algeria	100
14	13	109	Gazprom	Russia	50.023
15	15	113	Petrobras	Brazil	32.20
16	16	123	Rosneft	Russia	75.16
17	18	127	Lukoil	Russia	
18	17	129	Petronas	Malaysia	100
19	18	135	Adnoc	UAE	100

Appendix 1.1 Continued

2008 Rank	2007 Rank	PIW Index	Company	Country	State Ownership (%)*
20	21	141	Eni	Italy	30
21	20	152	NNPC	Nigeria	100
22	22	161	QP	Qatar	100
23	24	162	INOC†	Iraq	100
24	23	168	Libya NOC	Libya	100
25	25	173	Sinopec	China	75.84
26	27	175	EGPC	Egypt	100
27	26	184	StatoilHydro	Norway	65
28	28	188	Repsol YPF	Spain	
29	29	192	Surgutneftegas	Russia	
30	30	222	Pertamina	Indonesia	100
31	31	228	ONGC	India	74.14
32	32	236	Marathon	US	
33	32	258	PDO	Oman	60
34	34	265	EnCana	Canada	
35	34	266	Uzbekneftegas	Uzbekistan	100
36	36	269	Socar	Azerbaijan	100
37	43	279	TNK-BP‡	Russia	
38	39	295	Apache	US	
38	39	295	CNR	Canada	
40	37	297	SPC	Syria	100
41	50	300	Kazmunaigas	Kazakhstan	100
42	42	301	Devon Energy	US	
42	45	301	Hess	US	
44	41	302	Anadarko	US	
44	46	302	Occidental	US	
44	44	302	OMV	Austria	31.05
47	47	306	BG	UK	
48	48	314	CNOOC	China	66.41
49	52	317	Novatek	Russia	
50	38	325	Ecopetrol	Colombia	89.90

Source: Petroleum Intelligence 2009.

Appendix 2.2 Existing and Forthcoming Nuclear Reactors

Country	Reactors Operable		Reactors under Construction		Reactors Planned		Reactors Proposed	
	No.	*MWe net*	*No.*	*MWe gross*	*No.*	*MWe gross*	*No.*	*MWe gross*
Argentina	2	935	1	745	2	773	1	740
Armenia	1	376	0	0	1	1060		
Bangladesh	0	0	0	0	0	0	2	2000
Belarus	0	0	0	0	2	2000	2	2000
Belgium	7	5943	0	0	0	0	0	0
Brazil	2	1901	1	1405	0	0	4	4000
Bulgaria	2	1906	0	0	2	1900	0	0
Canada	18	12679	2	1600	4	4400	3	3800
China	13	10234	23	25900	39	44270	120	120000
Czech Republic	6	3686	0	0	2	2400	1	1200
Egypt	0	0	0	0	1	1000	1	1000
Finland	4	2721	1	1650	0	0	2	3000
France	58	63236	1	1650	1	1630	1	1630
Germany	17	20339	0	0	0	0	0	0
Hungary	4	1880	0	0	0	0	2	2200
India	19	4183	4	2720	20	16740	40	49000
Indonesia	0	0	0	0	2	2000	4	4000
Iran	0	0	1	1000	2	2000	1	300
Israel	0	0	0	0	0	0	1	1200
Italy	0	0	0	0	0	0	10	17000
Japan	55	47348	2	2756	12	16538	1	1300
Jordan	0	0	0	0	1	1000		
Kazakhstan	0	0	0	0	2	600	2	600
Korea DPR (North)	0	0	0	0	0	0	1	950

Appendix 2.2 Continued

Country	Reactors Operable		Reactors under Construction		Reactors Planned		Reactors Proposed	
	No.	*MWe net*	*No.*	*MWe gross*	*No.*	*MWe gross*	*No.*	*MWe gross*
Korea RO (South)	20	17716	6	7000	6	8400	0	0
Lithuania	0	0	0	0	0	0	1	1700
Malaysia	0	0	0	0	0	0	1	1200
Mexico	2	1310	0	0	0	0	2	2000
Netherlands	1	485	0	0	0	0	1	1000
Pakistan	2	400	1	300	2	600	2	2000
Poland	0	0	0	0	6	6000	0	0
Romania	2	1310	0	0	2	1310	1	655
Russia	32	23084	10	8960	14	16000	30	28000
Slovakia	4	1760	2	880	0	0	1	1200
Slovenia	1	696	0	0	0	0	1	1000
South Africa	2	1800	0	0	0	0	15	2000
Spain	8	7448	0	0	0	0	0	0
Sweden	10	9399	0	0	0	0	0	0
Switzerland	5	3252	0	0	0	0	3	4000
Thailand	0	0	0	0	2	2000	4	4000
Turkey	0	0	0	0	4	4800	4	5600
Ukraine	15	13168	0	0	2	1900	20	27000
UAE	0	0	0	0	4	5600	10	14400
United Kingdom	19	10962	0	0	4	6680	9	12000
USA	104	101229	1	1218	9	11800	22	31000
Vietnam	0	0	0	0	4	4000	10	11000
Total World	441	376,313	58	60,484	152	167,401	337	382,825

Source: World Nuclear Association 2010.

References

Adams, R.N. 1982. *Paradoxical Harvest: Energy and Explanation in British History, 1870–1914*. Cambridge: Cambridge University Press.

Andrews, J. and Jelley, N. 2007. *Energy Science: Principles, Technologies, and Impacts*. Oxford: Oxford University Press.

Asif, M. and Muneer, T. 2007. Energy supply, its demand and security issues for developed and emerging economies. *Renewable and Sustainable Energy Reviews*, 11(7), 1388–413.

Bauen, A. 2006. Future energy sources and systems – Acting on climate change and energy security. *Journal of Power Sources*, 157(2), 893–901.

Bickel, J.E. and Lane, L. 2010. Climate engineering, in *Smart Solutions to Climate Change: Comparing Costs and Benefits*, edited by B. Lomborg. Cambridge: Cambridge University Press, 9–52.

Bielecki, J. 2002. Energy security: Is the wolf at the door? *The Quarterly Review of Economics and Finance*, 42(2), 235–50.

Bilgin, M. 2007. New prospects in political economy of inner-caspian hydrocarbons and Western energy corridor through Turkey. *Energy Policy*, 35(12), 6383–94.

Bilgin, M. 2009. Geopolitics of European natural gas demand: Supplies from Russia, Caspian and the Middle East. *Energy Policy*, 37(11), 4482–91.

Bilgin, M. 2010. New energy order and FAST principles: Premises of equitable and sustainable energy security in the twenty-first century. *International Journal of Global Energy Issues*, 33(1–2), 4–22.

Blair, J.M. 1976. *The Control of Oil*. New York: Pantheon Books.

Bodansky, D. 2004. *Nuclear Energy: Principles, Practices, and Prospects*. New York: Springer.

CBF 2010. Mr. Five-per-cent Calouste Sarkis Gulbenkian. [Online: Calouste Gulbenkian Foundation]. Available at: http://www.gulbenkian.pt/section2 artId8langId2.html [accessed: 1 June 2010].

Chalvatzis, K.J. and Hooper, E. 2009. Energy security vs. climate change: Theoretical framework development and experience in selected EU electricity markets. *Renewable and Sustainable Energy Reviews*, 13(9), 2703–9.

Chew, E. and Barta, P. 2010. Malaysia Names New CEO for Petronas. *The Wall Street Journal*. [Online: 5 February]. Available at: http://online.wsj.com/article/SB10001424052748704259304575044340912128082.html [accessed: 28 October 2010].

Cook, E. 1971. The flow of energy in an industrial society. *Scientific American*, 224(3), 134–48.

Cordesman, A.H. 2003. *Saudi Arabia Enters the 21st Century*. Westport: Praeger.

Doyle, A. and Wynn, G. 2010. Copenhagen Accord climate pledges too weak. *Reuters*. [Online: 31 March]. Available at: http://www.reuters.com/article/idUSTRE62U13M20100331 [accessed: 29 September 2010].

Eckert, V. 2010. World 2009 CO2 emissions off 1.3 percent. *Reuters*. [Online: 13 August]. Available at: http://www.reuters.com/article/idUSTRE67C1I U20100813 [accessed: 28 September 2010].

EIA 2010. *International Energy Outlook 2010*. [Online: US Energy Information Administration, 27 July 2010]. Available at: http://www.eia.doe.gov/oiaf/ieo/index.html [accessed: 23 September 2010].

Falola, T. and Genova, A. 2005. *The Politics of the Global Oil Industry*. Westport: Praeger.

Fanchi, J.R. 2004. *Energy: Technology and Directions for the Future*. London: Elsevier.

Garg, A. and Shukla, P.R. 2009. Coal and energy security for India: Role of carbon dioxide (CO_2) capture and storage (CCS). *Energy*, 34(8), 1032–41.

Grant, A. 1888. *Force and Energy: A Theory of Dynamics*. London: Longmans, Green and Co.

Hoffmann, P. 2002. *Tomorrow's Energy: Hydrogen, Fuel Cells, and the Prospects for a Cleaner Planet*. Massachusetts: MIT Press.

Holland, G. and Provenzano, J.J. 2007. *The Hydrogen Age: Empowering a Clean-Energy Future*. Utah: Gibbs Smith.

Holt, M. 2009. *Nuclear Energy Policy.* CRS Report for Congress. Washington: Congressional Research Service.

Hore-Lacy, I. 2006. *Nuclear Energy in the 21st Century*. London: World Nuclear University Press.

Howard, R. 2004. *Iran in Crisis? Nuclear Ambitions and the American Response*. London: Zed Books.

Hoyos, C. 2007. The new Seven Sisters: Oil and gas giants dwarf western rivals. *Financial Times*. [Online: 11 March]. Available at: http://www.ft.com/cms/s/2/471ae1b8-d001-11db-94cb-000b5df10621.html [accessed: 24 November 2009].

IDC 2010. Historical Crude Oil Prices. [Online: Inflationdatacom, 21 July 2010]. Available at: http://www.inflationdata.com/inflation/inflation_rate/historical_oil_prices_table.asp [accessed: 13 October 2010].

Kong, B. 2010. *China's International Petroleum Policy*. Santa Barbara: ABC-CLIO-LLC.

Kozloff, N. and Chávez, H. 2007. *Oil, Politics, and the Challenge to the US*. New York: Palgrave Macmillan.

Kumar, B., Hiremath, R.B., Balachandra, P. and Ravindranath, N.H. 2009. Bioenergy and food security: Indian context. *Energy for Sustainable Development*, 13(4), 265–70.

Lefèvre, N. 2010. Measuring the energy security implications of fossil fuel resource concentration. *Energy Policy*, 38(4), 1635–44.

Makarieva, A.M., Gorshkov, V.G. and Li, B.L. 2008. Energy budget of the biosphere and civilization: Rethinking environmental security of global renewable and non-renewable resources. *Ecological Complexity*, 5(4), 281–8.

Maugeri, L. 2006. *The Age of Oil: The Mythology, History, and Future of the World's Most Controversial Resource.* Westport: Praeger.

Melosi, M. 2006. Energy transitions in historical perspective, in *Energy and Culture*, edited by B. Dooley. Aldershot: Ashgate, 3–19.

Morse, E.L. and Jaffe, A.M. et al. 2001. *Strategic Energy Policy Challenges for the 21st Century.* [Online: Report of an Independent Task Force, James A. Baker III Institute for Public Policy of Rice University and the Council on Foreign Relations]. Available at: http://www.rice.edu/energy/publications/docs/TaskForceReport_Final.pdf [accessed: 23 September 2010].

Msangi, S., Ewing, M., Rosegrant, M.W. and Zhu, T. 2010. Biofuels, food security, and the rnvironment: A 2020/2050 perspective, in *Global Change: Impacts on Water and Food Security*, edited by C. Ringler, A.K. Biswas and S. Cline. New York: Springer, 65–94.

Nersesian, R.L. 2007. *Energy for the 21st Century.* New York: M.E. Sharpe.

Nolan, P., Shipman, A. and Rui, H. 2004. Coal liquefaction, Shenhua Group, and China's energy security. *European Management Journal*, 22(2), 150–64.

Nuttall, W.J. and Manz, D.L. 2008. A new energy security paradigm for the twenty-first century. *Technological Forecasting and Social Change*, 75(8), 1247–59.

OECD 2010. Member and Non-member Countries. Available at: http://www.oecd.org/countrieslist/0,3351,en_33873108_33844430_1_1_1_1_1,00.html [accessed: 24 September 2010].

O'Neill, J. 2001. Building better global economic BRICs. *Goldman Sachs Global Economics Paper*, 66 (30 November). Available at: http://www2.goldmansachs.com/ideas/brics/building-better-doc.pdf [accessed: 24 September 2010].

OPEC 2009. *Brief History.* Available at: http://www.opec.org/aboutus/history/history.htm [accessed: 1 June 2010].

Parra, F. 2005. *Oil Politics: A Modern History of Petroleum.* London: I.B. Tauris.

PIW 2009. PIW's top 50: How the firms stack up. *Petroleum Intelligence Weekly.* [Online: 30 November]. Available at: http://www.energyintel.com/DocumentDetail.asp?document_id=648479 [accessed: 23 September 2010].

Pidd, H. 2010. BP oil spill estimates double. *The Guardian.* [Online: 11 June]. http://www.guardian.co.uk/environment/2010/jun/11/bp-oil-spill-estimates-double [accessed: 30 July 2010].

Quaschning, V. 2010. *Renewable Energy and Climate Change.* Chichester: John Wiley & Sons.

Rosner, K. 2006. *Gazprom and the Russian State.* London: GMB Publishing.

Salameh, M.G. 2003. The new frontiers for the United States energy security in the 21st century. *Applied Energy*, 76(1–3), 135–44.

Slater, R. 2010. *Seizing Power: The Grab for Global Oil Wealth.* New Jersey: John Wiley & Sons.

Spero, J.E. and Hart, J.A. 2009. *The Politics of International Economic Relations.* Boston: Wadsworth Cengage.

Stern, R. 2007. The Iranian petroleum crisis and United States national security. *PNAS*, 104(1), 377–82.

Thomas, V.M., Choi, D.G., Luo, D., Okwo, A. and Wang, J.H. 2009. Relation of biofuel to bioelectricity and agriculture: Food security, fuel security, and reducing greenhouse emissions. *Chemical Engineering Research and Design*, 87(9), 1140–46.

UN 1992. *Report of the United Nations Conference on Environment and Development*. Rio de Janeiro, 3–14 June, A/CONF.151/26 (Vol. I). [Online: 12 August 1992]. Available at: http://www.un.org/documents/ga/conf151/aconf15126-1annex1.htm [accessed: 29 September 2010].

UN 1998. *Kyoto Protocol to the United Nations Framework Convention on Climate Change*. Available at: http://unfccc.int/resource/docs/convkp/kpeng.pdf [accessed: 29 September 2010].

Vermerris, W. 2008. Why bioenergy makes sense, in *Genetic Improvement of Bioenergy Crops*, edited by W. Vermerris. New York: Springer, 3–42.

Weissenbacher, M. 2009. *Sources of Power: How Energy Forges Human History*. Santa Barbara: Greenwood.

WNA 2010. *World Nuclear Power Reactors and Uranium Requirements*. [Online: World Nuclear Association, 1 October]. Available at: http://world-nuclear.org/info/reactors.html [accessed: 6 October 2010].

Youngs, R. 2009. *Energy Security: Europe's New Foreign Policy Challenge*. New York: Routledge.

Chapter 3
A Grounded Comparison of Energy Security in Denmark, Brazil, Bangladesh, and China

Benjamin K. Sovacool

Introduction

With energy services instrumental to modern economies and postmodern lifestyles, energy security is becoming paramount to human security. Coal, oil, natural gas, and uranium are currently needed to energize our factories, light schools and workplaces, produce food, power our vehicles, and cool and warm our residences. The late economist E.F. Schumacher (1977) once ruminated that energy was "not just another commodity, but the precondition of all commodities, a basic factor equal with air, water, and earth." Yet because of its ubiquitous nature, the benefits of improved energy security are often non-rivalrous and non-excludable, similar to other public goods like national defense or clean air. As Bielecki (2002: 236) opines:

> energy security is a public good which is not properly valued by the market and the benefits of which are available equally to those who pay for it and to those who do not. Consequently, the market may tend to produce a level of energy security that is less than optimal.

Thus a paradox arises: energy security is integral to modern society, yet its very ubiquity makes it prone to market failure and under-distribution. How, then, can policies and programs be designed to incentivize energy security? Drawing from research interviews and field research in seven countries along with four case studies, this chapter explores the conditions that result in programs and policies that improve national energy security. After explaining its case selection and research methods, the study explores four case studies related to electricity supply in Denmark, ethanol production in Brazil, small-scale renewable energy in Bangladesh, and off-grid energy use in China. It concludes that programs that foster equity, inclusivity, information, incentives, organizational multiplicity, and adaptability successfully improve various elements of energy security.

Case Selection and Research Methods

Energy security, simply put, involves how to equitably provide available, affordable, reliable, efficient, environmentally benign, properly governed and socially acceptable energy services (Sovacool 2010; Brown and Sovacool 2011). In searching for representative case studies of energy security, the author chose to focus on four areas: electricity supply, transport, household energy use, and offgrid energy services. The author chose electricity supply because electricity accounts for well more than half of global energy consumption each year; transportation because it comes second after electricity in total energy use (International Energy Agency 2010). The author selected Denmark for electricity supply because of its high rate of renewable energy diversification, and Brazil since it, along with the United States, dominates the global ethanol market. Household energy use and offgrid energy services were chosen because, in contrast to electricity and petroleum-based motorized transport in industrialized countries, 3 billion people still rely on traditional biomass fuels for cooking and heating, 1.5 billion have no access to electricity, and an additional 1 billion have access to only unreliable electricity networks (United Nations Development Program 2010). The author therefore also selected Bangladesh for its program promoting solar home systems and biogas units, and China for its efforts to distribute improved cook stoves.

In selecting a methodology, the analysis relies on a case study, grounded, ethnographic, semi-structured interview approach described in Appendix A. Forty-five semi-structured research interviews were conducted with experts in seven countries at 27 institutions over from October 2007 to July 2009. Site visits to projects in Australia, Brazil, Denmark, Germany, India, Indonesia, Malaysia, Singapore, Spain, the United Kingdom, and the United States (some of these visits ended up excluding case studies) were made. When arranging interviews, special care was taken to include:

- Universities and academic institutions such as the University of California-Berkeley, the University of Delhi, School of Public Health at Rutgers University, and the University of São Paulo;
- Nongovernmental organizations such as the International Energy Foundation and Forest Stewardship Council;
- Research institutes such as the Sugarcane Technology Center in Brazil and Risø National Laboratory in Denmark;
- Regulatory agencies and government ministries;
- Companies and manufacturers such as Vestas and Dong Energy.

Respondents were asked to discuss the history, challenges, and benefits to each case, and also to recommend relevant literature confirming their points (with much of this academic literature cited below). Although these interviews have been carefully transcribed for accuracy, to encourage candor and protect confidentiality, the author has presented such data as anonymous. Unless otherwise indicated, all of the data for each of the case studies comes from these research interviews, although it is impossible to ascribe particular comments and views to specific individuals. More details about these interviews are presented in Appendix B.

This sections that follow explore four effective approaches to improving energy security: electricity policy in Denmark, ethanol production in Brazil, Grameen Shakti in Bangladesh, and off-grid energy use in China. Each section begins with a summary of the case before tracing its historical genesis, explaining the policy mechanisms and technologies involved, and identifying its contribution towards fighting climate change and/or improving energy security. The final section teases out a set of common elements within the case studies to better inform energy policy and planning.

Danish Electricity Policy, 1970–2001

Denmark has relied upon a progression and variety of policy mechanisms at different scales and involving different actors to reduce greenhouse gas emissions, including taxes, research subsidies, and feed-in tariffs to promote energy efficiency, combined heat and power, and wind energy. To list just some of the most significant achievements, Denmark transitioned from being almost 100 percent dependent on imported fuels such as oil and coal for their power plants in 1970 to becoming a net exporter oil and electricity today. The country leads the world in terms of exportation of wind energy technology, with a hold on roughly one-third of the world market for wind turbines. Denmark has the largest portfolio of wind projects integrated into their power grid of any country in the world (more than 20 percent), and some parts of the country, such as Western Denmark, frequently supply more than 40 percent of their electricity from wind turbines. As more renewable energy and cogeneration units have come to replace less efficient and more polluting conventional fossil fuel units, total CO_2 emissions associated with fuel combustion and electricity generation have plummeted. In 1990, almost 1,000 grams of carbon dioxide were emitted for every kWh of electricity produced in Denmark. In 2005, less than 600 grams were emitted. Altogether, the carbon dioxide emission intensity – the amount of CO_2 emitted per unit of Gross Domestic Product – was 48 percent lower in 2004 than it was in 1980. These achievements are all the more impressive when readers consider that the country is roughly the same size and population as the state of Maryland in the United States.

At the core of Denmark's successful approach to electricity policy has been sustained taxes on energy fuels, electricity, and carbon dioxide, which both create incentives for energy efficiency and provide government revenue for renewable energy research programs. Higher gasoline, diesel, and oil taxes were first passed in 1974 following the oil shocks from the OPEC embargo. These measures were followed by additional taxes for coal in 1982, carbon dioxide in 1992, and natural gas and sulfur in 1996. These taxes raised more almost $9 billion in 2005 and, cumulatively, $25 billion from 1980 to 2005.

These energy related taxes were kept high and not lowered after fossil fuel prices dropped in the 1980s and 1990s so that the manufacturing industry could rely on stable fuel and electricity prices. The taxes furthermore sent price signals that encouraged energy efficiency measures in the Danish power market, and accrued government funds for R&D expenditures that they directed at wind power, biomass, and small-scale combined heat and power units. Primary energy consumption nationally has grown just four percent from 1980 to 2004, even though the economy grew more than 64 percent in fixed prices. The carbon tax alone added about 1.3 euro cents per kWh of additional income for renewable power providers.

The energy taxes also helped fund a government program prompted by researchers and local communities interested in using low-tech windmill designs to generate electricity. Danish regulators, working with manufacturers and interested citizens, blended a bottom-up strategy of wind turbine development – a slow, crafts oriented, step-by-step process including incremental learning through practical experience – with national regulation. The Danish model adhered to what one respondent called "learning by doing;" and designers recognized the learning curve needed to perfect development and were willing to tolerate and adapt to earlier setbacks. Their approach paid dividends. According to researchers at the Risø National Laboratory, from 1980 to 2005 the cost per kWh of Danish wind turbines decreased 70 percent, and Danish R&D enabled wind turbines to produce 180 times more electricity at 20 percent the cost. Over the same period, commercial turbine output grew 100 fold, from 30 kW in the 1980s to 3 MW in 2006.

As Danish wind technology matured, the government supported its expansion with additional policies. Starting in 1979, the Danish government promoted an investment subsidy that reimbursed individuals, municipalities, and farming communities for the capital cost of installing wind, solar, and biogas digesters. These subsidies initially covered 30 percent of expense of renewable energy systems but were scaled down periodically as the industry matured and turbine prices decreased. Just two years later in 1981, the government passed a feed-in tariff requiring utilities to buy all power produced from renewable energy technologies at a rate equal to 70 to 85 percent the consumer retail price of electricity in a given distribution area (Morthorst 2000). In 1985, an agreement was reached between the government and the electricity utilities, committing the utilities to install 100 MW over a five-year period (and fully implemented by 1992).

That same year, lawmakers passed two other important policies. The Danish government established the "Danish Wind Turbine Guarantee." This guarantee provided long-term financing of large wind projects that used Danish made wind turbines, reducing the risk of building larger projects and encouraging local manufacturing. The Danish Energy Authority also provided open and guaranteed access to the grid. Grid connection costs were to be shared between the owner of the wind turbine and the electricity utility. The wind turbine owner had to bear the costs of the low voltage transformer and connection to the nearest connection point on the distribution grid. The utilities had to cover the costs for reinforcement of the distribution grid when needed.

As a consequence of these policies, Danish Transmission System Operators were legally obligated to finance, construct, and operate the transformer stations and transmission and distribution infrastructure for centralized wind farms and decentralized wind turbines owned by ordinary people. They were required to connect wind power and expand the grid if necessary, and provide financial compensation if any of the wind power generated was curtailed. The costs of this infrastructure investment and reimbursement for curtailed power were paid for by the government, and then distributed to all customers.

To accommodate the technical difficulty of managing a highly dispersed and decentralized electricity system, Danish transmission networks are comprised of mostly newer equipment with a preponderance of high voltage transmission lines delivering power over smaller distances with low voltage levels. A significant number of transmission lines are underground, and system operators report no unaccounted loss – in other words, the grid is so "tight" that every kWh is counted. This "tightness" is needed to trade power between its Nordic neighbors in the north and Germany in the south, and it is bolstered by the large hydroelectric and pumped hydroelectric reserves of Finland, Norway, and Sweden.

Another key component of Danish electricity policy has been the aggressive support for combined heat and power and cogeneration power plants. Quickly after the government abolished the installation subsidy for wind energy in 1989, it promoted environmentally friendly zoning in 1990 to advance electricity investment for towns and villages outside major cities. Cogeneration units were required to replace district heating units, and their previous use of oil, diesel, and coal was prohibited and replaced by natural gas. If the local market was not large enough to cater cogeneration, the district heating plants were required to utilize biomass. Concomitantly, all large utilities in the major cities were ordered to use biomass (especially straw), and required to obey mandatory energy efficiency regulations. This wave of environmentally friendly conversions and improvement of efficiency drove significant investment in the combined heat and power market. Thus, cogeneration now provides roughly 60 percent of all electricity and 80 percent of heat consumed in Denmark, making it a leader in Europe. The next closest country, the Netherlands, harnesses only 38 percent for electricity, Finland 36 percent, and every other country in the European Union less than 10 percent.

The greater penetration of cogeneration and wind energy has promoted the decentralization of energy supply, a transition that has brought political benefits (in the form of greater diversification) and energy security benefits (by minimizing the number of vulnerable power plants on the grid). Unlike the US and other countries, where the production and consumption aspects of energy are segregated (and power plants frequently "out of sight, out of mind") energy production in

Denmark is predominately decentralized and close to the end user. The country shifted from centralized generation (with less than 20 large-scale plants) in the 1970s to a decentralized model including more than 4,000 small-scale generators today. Even in the capital, Copenhagen, wind turbines are fully integrated into the landscape. They are in public parks, near factories, in the harbor, and inherently *visible*, meaning that people tend to view them as natural, instead of the US, where energy technologies are predominately *invisible*. There is such strong support for renewables in Denmark that when the government announced its most recent renewable energy targets, they were promptly criticized by numerous political parties for not being progressive enough.

Brazil's National Ethanol Program, 1975–1988

Brazil created its *Proálcool* program in November 1975 to increase ethanol production and substitute ethanol for petroleum in conventional vehicles. Brazil not only surpassed the initial goals of their program in its first three years, but emerged to become the largest producer of ethanol in the world, making 26.9 billion liters of ethanol in the 2008–2009 season. Annual ethanol production for Brazil now involves revenues of $20 billion and 7 million hectares of planted land, more than 72,000 farmers and sugarcane producers, 390 sugarcane mills, 240 distillery plants, 3.5 million direct and indirect jobs, and a harvest of 569 million tons of sugarcane, 106 million tons of bagasse (a byproduct used to make electricity), and 30 million tons of sugar. The Brazilian program was so successful that the government no longer needs to directly subsidize ethanol. During the period of the *Proálcool* program, fermentation times for ethanol production have been cut in half, yields have almost tripled, and efficiencies for sugarcane extraction, fermentation, distillation, and electricity generation have steadily improved.

While Brazilian researchers experimented with different blends of ethanol in the 1920s, an aggressive program for ethanol did not really emerge until the 1970s. The Brazilian government initiated a coordinated program to promote ethanol production in 1975 in response to the 1973 oil crisis (Goldemberg 2006; Goldemberg 2008). The *Programa Nacional do* Álcool, or *Proálcool*, was created after high gasoline prices resulted from the oil embargo of the Organization of Petroleum Exporting Countries. Brazilian regulators hoped that ethanol development would insulate the economy from the vagaries of the global market. *Proálcool* had two primary components: a compulsory mandate that all gasoline had to have at least a 10 percent blend of ethanol, and a voluntary mandate that automobile manufacturers design vehicles to run on 100 percent ethanol.

The first phase of *Proálcool* from 1975 to 1979 involved using annexed distilleries (distilleries attached to sugarcane mills) to produce anhydrous ethanol from molasses to be blended with gasoline in various mixes from five to 25 percent. Ethanol was used in cars that ran on blends of ethanol with gasoline, often called "gasohol," instead of pure ethanol, and the program was focused almost exclusively on growing sugarcane for ethanol production and adapting existing plantations and factories to produce ethanol. The blending of ethanol started slowly, beginning in Sao Paulo and the Northeast before expanding to the rest of the country. Researchers started focusing on an engine that could operate entirely on ethanol (called a "neat" ethanol engine), but left it to partnerships with automobile manufacturers to produce results. Under this first phase of the program, ethanol production increased tenfold from 220 million liters in 1975 to 2.8 billion liters in 1979.

The second phase of the program from 1979 to 1984 focused on expanding ethanol production and increasing the efficiency of the ethanol distillation process. Independent sugarcane distilleries started to emerge no longer coupled to sugar mills, and these newly autonomous distilleries produced anhydrous ethanol (for blending) and hydrous ethanol (pure) from both sugar and molasses. To use hydrous ethanol, regulators also shifted the program from trying to promote gasohol to promoting neat ethanol vehicles, which started entering the commercial market in 1980. At this time, the government also regulated that the pump price of hydrous ethanol would be equivalent to 64.5 percent of the price of gasoline and used higher gasoline taxes to pay for any true differences in price between the two fuels. This mean that pure ethanol was always cheaper than its alternatives; that purchases of gasoline actually helped fund the ethanol program; and that drivers were incentivized to purchase neat-only ethanol vehicles, whose sales jumped so dramatically that in 1981 they accounted for more than 90 percent of all new vehicle sales.

When public mistrust and suspicion started to occur in 1981 over high ethanol prices and poor performance of ethanol vehicles, the government worked with auto manufacturers to improve ethanol powered engines, set a price cap on ethanol to stabilize prices, and ran an educational program to inform consumers. By 1983, consumer confidence in ethanol resumed and the sale of ethanol-fueled vehicles reached 84 percent of total vehicle sales in Brazil. By the end of the second phase of the program in 1984, the total fleet of neat ethanol vehicles reached five million and most service stations sold exclusively ethanol or ethanol and gasohol.

The government protected sugarcane workers by giving them additional bonuses in salary related to value of sugar and ethanol produced each year. While working conditions on sugar plantations and ethanol distilleries were indeed strenuous, workers in Sao Paulo received wages more than 80 percent higher than the agricultural sector average. Environmental groups were invited to solicit feedback and resulted in the government mandating that rainforested land could not be used for new sugarcane plantations, processing facilities, or ethanol plants. Brazilian automakers were enrolled early on, and reached an agreement with the government in 1979 to produce cars that ran first on blends of ethanol and gasoline and then purely on ethanol. This agreement with all major domestic manufacturers resulted in the production of 250,000 ethanol-only vehicles by 1980 and 350,000 by 1982 (Cordonnier 2008). Later partnerships were made with international automobile companies such as Ford, Fiat, and Volkswagen. The government sought the involvement of gasoline distributors, providing them with low-cost pumps and tanks for ethanol. As ethanol production increased partnerships were cemented between ethanol producers and electricity suppliers, since ethanol made from sugarcane produces bagasse, a usable fuel for power pants.

While the program underwent structural changes after 1985, it cost about 20 cents to produce a liter of ethanol using Brazilian sugarcane in 2007 but 47 cents to produce a liter of ethanol from corn and 97 cents to produce a liter from cereals. For Brazil, these benefits translate into real savings at the pump. In 2008, a liter of ethanol at a typical gasoline station in Brazil cost less than *half* a liter of gasoline. The ethanol program has also greatly reduced Brazil's dependence on imported oil with substantial macroeconomic savings. In 1970 before *Proálcool* started Brazil was dependent on foreign suppliers for 80 percent of its oil supply, yet in 2009 more than 60 percent of gasoline was substituted by ethanol. While the net cost to the central government in terms of subsidies for *Proálcool* was about $11 billion from 1975 to 1985, the nation saved $55 billion in avoided oil imports over the same period. From 1975 to 2000, the entire ethanol program cost the government about $30 billion but saved them that amount every two years. Confirming that the benefits from ethanol expand beyond merely displaced oil imports, two independent economists not affiliated with the Brazilian program estimated that the country's external debt was $100 billion less than it would have been in the absence of ethanol production (Childs and Bradley 2007: 26–7).

Another important benefit from Brazil's ethanol program has been electricity generation from bagasse. Sugarcane cultivation produces a byproduct know as bagasse, the fibrous part of the sugarcane plant that remains after its stalks have been crushed and their juice extracted. Bagasse is such a versatile material that it has even been used to make disposable diapers and greeting cards. One key component of the Brazilian program is that virtually all of the energy needs for the production process – the electricity required to crush, ferment, distill, and refine sugarcane into ethanol – are met by onsite, decentralized bagasse-fired electricity generators. In 2008, the ethanol industry produced so much bagasse that they met all of their energy needs *and* exported additional electricity to the grid at a low cost of 6.5 cents per kWh. While this price was still above the cheap electricity coming from Brazil's vast network of hydroelectric dams, it was competitive enough so that bagasse-fired electricity displaced the need to operate 3,000 MW of installed fossil-fueled capacity. Making electricity from bagasse is becoming so profitable that while the typical sugar mill today receives 56 percent of its income from sugar, 43 percent from ethanol, and one percent from electricity, the industry expects ethanol to grow to a 54 percent share and electricity to a 16 percent share by 2025.

As a perhaps odd benefit given the recent controversy over connection between biofuel crop production and deforestation, sugarcane cultivation can actually improve soil fertility and restore degraded land. Unlike wheat and soy which usually need irrigated water and fertilizer to grow, sugarcane is a grass and can grow by itself in Brazil without the need for external inputs. When the right variety if planted in the right place, sugarcane cane can maintain soil fertility and actually improve it in formerly degraded land areas, create habitats for species and insects, and prevent runoff and sedimentation. Sugarcane acts as a carbon sink since carbon dioxide is sequestered in its roots (which are not uprooted during harvesting). Each hectare of sugarcane removes 10 to 28 tons of carbon dioxide pear year while the same space with corn removes only 2 to 4 tons. Planters in Brazil also rely on a rich variation of sugarcane types to minimize the risks associated with monocultures. Each plantation has only 5 to 7 percent of its land area covered with one variety of sugarcane, a practice that promotes diversity and traces back to indigenous farming practices.

Because of these attributes – greater energy content, byproducts such as bagasse, local production and farming – Brazilian ethanol has a much lower carbon and pollution footprint than gasoline, diesel, and other biofuels. While sugarcane harvesting can release particulate matter (a dangerous pollutant) into the air when done through burning, lifecycle greenhouse gas emissions are much lower for sugarcane ethanol than gasoline – with ethanol releasing 0.6 kilograms of carbon dioxide per liter compared to one kilogram for gasoline (Leite et al. 2009). When put into a national context, about 27.5 million metric tons of greenhouse gas emissions were avoided in 2008 due to the use of ethanol replacing gasoline in Brazil and an additional 5.7 million tons were avoided through the use of bagasse-fired electricity. From 1970 to 2008 about 806 million metric tons of carbon dioxide equivalent have been avoided from the use of ethanol all together, an amount greater than the emissions from *all* of Oceana (Australia, Fiji, New Zealand, Papua New Guinea, and the Solomon Islands) for one year.

Grameen Shakti in Bangladesh, 1996–2009

The nonprofit company Grameen Shakti (GS) provides financial and technical assistance for renewable energy projects in rural Bangladesh. Based on experiences from the Grameen Bank's microcredit program, GS relies on a market-based approach to promote solar home systems, small-scale biogas plants, and improved cook stoves to reduce deforestation, fight poverty, and provide energy services. Other novel factors include a focus on matching energy supply with income generating activities, relying on local knowledge and entrepreneurship, utilizing community awareness campaigns, and innovative payment methods including fertilizer, livestock, and cash. As of July 2009, GS operated 750 offices throughout every district of Bangladesh, had installed 250,000 solar home systems, 40,000 cook stoves, and 7,000 biogas plants among 2.5 million recipients. They plan to ramp up their expansion so that by 2012 one million solar home systems are in place along with half a million biogas plants and 10 million improved cook stoves.

Inspired by the success of the Grameen Bank, a specialized bank that gives small-scale loans to poor women (referred to as "microcredit") who also own nearly 98 percent of the bank, Grameen Shakti was established in June 1996 to distribute renewable energy systems to the rural population. The organization was created to explicitly promote knowledge and awareness about renewable energy, provide technical training on solar energy to the rural workforce, and overcome the high upfront cost of installing solar and biogas systems. Taken from the Sanskrit word for "energy" or "empowerment," GS (which literally means "village energy" or "village empowerment") was kicked off with a $750,000 loan from the International Finance Corporation as well as a number of small grants, although it quickly repaid these loans and became a self sustainable nonprofit company. As of March 2009, the organization had 12 divisional offices, 94 regional offices, and a total number of 5,000 employees, mostly engineers. It received a small amount of funding from the Infrastructure Development Company Limited (a government organization in Bangladesh that finances rural based renewable energy companies) but otherwise received all of its money from sales. Three of their programs – solar home systems, biogas, and improved cook stoves – have been the most effective.

The solar home systems (SHS) program draws on the decentralized nature of solar photovoltaic panels to provide electricity to off-grid and inaccessible areas in rural Bangladesh. The SHS program targets those areas that have little to no access to electricity and limited opportunities to become connected to centralized electricity supply within the next five to ten years. The ease of operating solar panels, their long lifespan, avoidance of combustible fuel, and lack of pollution make them an ideal choice for remote areas, and the SHS offers microcredit schemes to enable homeowners and businesses to acquire the necessary capital they need to finance installation.

Under the SHS program interested parties make a down payment to cover 15 to 25 percent of the system cost and then repay GS with a low-interest loan over two to three years. Given the high costs of kerosene and diesel in rural parts of Bangladesh, systems typically pay for themselves in three to four years, meaning people that purchase them then own a system that lasts 20 years without fuel costs. Customers can also elect to share in the cost of larger systems under a solar micro-utility scheme that allows shopkeepers or villages to share in making a 10 percent down payment spread across 42 months of repayment. GS offers free maintenance while their loans are being repaid, and trains interested clients in maintenance and operation at no additional cost. As of 2009, GS has installed more than 250,000 solar home systems with a production capacity of 50 MWh-peak to more than two million people. About 10,000 clients have availed themselves of the micro-utility financing option, and the program is growing at a rate of 10,000 new clients per month.

To keep on top of maintenance and the expansion of the SHS program, GS has established 45 Grameen Technology Centers (GTCs) that have trained more than 40,000 female heads of households in the proper use and maintenance of solar panels. GS engineers pay monthly visits to households during their financing period, and they offer their services for a small fee, afterwards, if a client signs an annual maintenance agreement. The GTCs have trained more than 5,000 women in advanced solar maintenance, enabling them to become full-time specialists who travel around the country to service GS clients. Apart from providing repair and maintenance services, GTC trained women technicians also assemble solar accessories such as lamp shades, charge controllers, and inverters, and GTCs run exposure programs for rural school children to increase their awareness about renewable energy.

The GS biogas program promotes small-scale, three cubic meter biogas plants to be used in homes and residences, suitable for providing gas and heat for cooking. The biogas program can be subscribed to at the commercial scale as well, with larger systems offering enough gas to meet the energy needs of restaurants, tea stalls, and bakeries. By relying on biogas, these units minimize reliance on traditional forms of biomass, animal dung, and charcoal (with their negative environmental and social impacts), and also protect communities from disease by enhancing sanitation. The plants harness gas obtained from livestock and, yes, even human excrement. Researchers estimated that Bangladesh had more than 30 billion cubic meters of potential biogas from livestock and another 10 billion cubic meters from human beings (Islam et al. 2006). The biogas program quite literally has people using their own waste to meet their energy needs.

The financing scheme behind the biogas program is similar to the SHS. Purchasers pay for 25 percent of the total cost of each system as a down payment, and then repay the rest in 24 monthly installments with a 6 percent service charge. Buyers are encouraged to construct their own plant under the supervision of GS engineers. Biogas plants at the community scale have proven to be quite effective, as many people in rural Bangladesh live in joined households with dozens of people close to each other, meaning they can easily share a biogas system. GS manages a special program in Chilmari (the northern part of the country) that provides farmers and communities with the livestock in addition to biogas plants so that they have adequate "fuel." More than 7,000 biogas units were installed by June 2009 and follow up evaluations have found that more than 90 percent of the plants installed under the project are still in operation and more than 90 percent of the households that use them can meet their fuel demand exclusively from these plants.

Lastly, the Improved Cook Stoves (ICS) program distributes one-, two-, and three-mouthed clay cook stoves which cut fuel use by half and have chimneys that create a smoke-free cooking environment. These more efficient cook stoves not only result in less fuel consumption (typically reducing fuel needs by 40 to 50 percent), they also facilitate shorter cooking times and generate and more heat. GS has recently started an ambitious program to install 10 million of these systems by 2012, and it has trained more than 2,000 local youths and women to manufacture, sell, and repair improved cook stoves. A special effort has been made to promote improved cook stoves among restaurants, soap manufacturers, and other food providers.

Collectively these programs have greatly reduced rates of deforestation. Bangladesh is a biomass-centered energy system, with trees and bamboo meeting about 48 percent of all domestic energy requirements followed by agricultural residues (36 percent) and dung (13 percent) (Miah et al. 2009). Widespread destruction of forests has occurred to satisfy energy needs, with homestead forest cover reduced to eight percent of its original area and half of Bangladesh's natural forests destroyed by people collecting fuel-wood in a single generation. Even when fuel collectors avoid cutting down trees and instead only remove fodder and kindling, they can still devastate forests by removing key nutrients from the ecosystem (Islam and Weil 2000). The programs operated by GS keep trunks, branches, shrubs, roots, twigs, leaves, and trees in the forest by minimizing the need for biomass and increasing the efficiency of cook stoves.

Another benefit relates to public health. Reliance on biomass combustion for cooking and lighting produces a significant quantity of hazardous pollutants, including fine particles, nitrogen oxide, and carbon monoxide, which are typically emitted within the home. One interview respondent compared this type of exposure to "living within a giant cigarette," and noted that homes reliant on biomass tended to have higher rates of acute respiratory infections, eye problems, low birth weights, and lung cancer. More than 50 percent of women in one survey of rural Bangladesh reported headaches, lung disease, asthma, *and* cardiovascular disease related to cooking with biomass fuels (Miah et al. 2009). Fuel collection also presents a health hazard, with many women carrying more than their weight in fuel hundreds of kilometers per month. These adverse health effects can lock families into poverty, as they increase expenditures for medical care while also diminishing productivity.

China's National Improved Cook Stove Program, 1983–1998

Despite a trend towards urbanization in recent years about 60 percent of the population in China is still rural, and 90 percent of all households in China still rely on solid fuels such as wood, biomass, and coal that they combust in their homes to meet their energy needs. To prevent shortages of these fuels among the rural population, minimize the degradation of forest lands and agricultural soils, and promote economic growth through a quick return on energy savings, the Chinese Ministry of Agriculture managed a National Improved Stove Program (NISP) from 1983 to 1998. Over this period, the NISP was responsible for the installation of 185 million improved cook stoves and facilitated the penetration of improved stoves from less than one percent of the Chinese market in 1982 to more than 80 percent by 1998. As a consequence, energy use per capita declined in rural areas at an annual rate of 5.6 percent from 1983 to 1990. (For excellent analysis on the history of the program readers should peruse Sinton et al. 2004; Qiu et al. 1996; and Smith et al. 1993).

While the rural population in China has grappled with population pressures, deforestation, and fuel shortages for more than a millennium, during the late 1970s and early 1980s millions of Chinese households experienced shortages of wood and fuel. Anecdotal stories flourished about peasants burning their furniture to keep warm, combusting uprooted plants and grasses to cook food, and chopping down bridges to create firewood. The Chinese Ministry of Agriculture started a National Improved Stove Program in 1983 to respond to these challenges under the direction of the State Development Planning Commission and the Ministry of Finance. While the characterization of phases is a bit rough due to the overlap between national and provincial programs, the first phase of the NISP ran from about 1983 to 1990, and it focused on the rapid dissemination of subsidies to three key stakeholders – county and provisional government officials, technical institutions, and households – to incentivize them to distribute safer and cleaner cook stoves. The Ministry of Agriculture formed the Bureau of Environmental Protection and Energy (BEPE) in 1983 to manage the program. BEPE adopted a "self building, self managing, self using" policy focused on having rural people themselves invent, distribute, and care for cook stoves, and BEPE set up pilot programs in hundreds of rural provinces to accomplish their policy.

The justification for embarking on a major cook stove program was related to the fact that most stoves in use at the time had efficiencies averaging 10 to 12 percent, meaning as much as 90 percent of the energy content of the wood or charcoal used in them was wasted. In some cases existing cook stoves could be drastically improved by something as simple as adding a chimney or more insulation around the stove to retain heat; in other cases older stoves could be replaced with new stoves with efficiencies greater than 20 to 30 percent, requiring half as much fuel and much less fuel collection. Newer stoves often required a switch away from charcoal or polluted wood to "healthier" fuels such soft biomass, crop residues, and firewood; they had a grate and an improved combustion chamber; and they almost always had a chimney. They utilized higher temperature ceramics, fire resistant material, longer lasting metals, had more insulation, and housed a better frame that guided hot gases up near the pot. They could cook more food at once and many had coils around the combustion chamber to heat water while cooking; some improved stoves were connected to radiators or space heaters so that heat could be recycled and/or vented to other rooms; and some stoves sent heat through pipes directly into a brick platform called a *kong* that occupants slept on at night. In rare occasions, improved stoves were "fuel flexible" and could combust coal and biomass, although doing so required a homeowners to insert a different combustion chamber for each fuel. Improved stoves were often aesthetically pleasing with beautifully designed tile and artwork, making them something to be proud of and handed down to children, something "treated as an asset rather than a consumable."

One key aspect of the national program is that it skipped many administrative levels of the Chinese government. BEPE disbursed subsidies directly from the central government to counties, meaning provincial officials had no say in how funds were spent, but it then enrolled provincial governments in monitoring and evaluating components. Under the first phase, BEPE signed 100 contracts with the rural counties most at need; these counties then received funding for three years through seven rounds of county selection. Counties were chosen on criteria including shortages of fuelwood, high reliance on woody biomass, and a local willingness to devote resources to the program. The key implementing agencies were County Rural Energy Offices (CREOs) at the local level.

Despite this somewhat complicated structure, the role of the national government was surprisingly limited. One primary aim of the NISP was to train local stove builders and installers so that they could carry out most of their work autonomously. Combustion chambers and ceramic tiles were often manufactured at the scale of villages (100 batches a day) instead of at huge industrial plants. Some builders specialized in stove construction while others retained unique knowledge about how to incorporate stoves into homes. CREOs often hired these local builders to design and install stoves and then sent demonstration teams to visit households personally to show villagers how to operate and maintain new stoves.

CREOs also exhibited a great variation in how they chose to spend the money given by BEPE to promote improved stoves. Officials in Wuhua County, for example, decided that only those households using an improved cook stove were allowed to cut wood in the mountains at a preferential price or could use the municipal woodpile, authorized the poorest households to obtain parts for their stoves without charge, awarded well performing townships with cash prizes and bonuses, penalized villager leaders that chose not to participate, and fined craftspersons for making traditional stoves. Yet in the more industrial Jiangjin County, officials gave special priority to local enterprises that actually manufactured stoves and focused efforts on installations in new houses. The NISP proved so successful in its early years that it became a centerpiece of the Seventh Five-Year Plan for 1986 to 1990, which called on using improved stoves (along with firewood forest plantations, small hydropower units, and biogas stoves) to deliver energy services to the rural Chinese. During this period, the total number of stoves installed jumped from 63.6 million to 120 million.

The second phase of the program occurred from 1990 to 1995. With shortages of biomass and woody fuels alleviated in many counties and rural incomes rising, the government started scaling back both national and provincial subsidies for the program. Officials made a major push for efforts related to commercialization, certification, and standardization. Regulators wielded tax breaks and low-interest loans to incentivize rural energy companies to become the primary distributors of stoves, although the government retained the responsibility for training and the provision of technical support. This phase also saw a shift in the development of stove models from those manufactured onsite to more industrial units that could be mass produced and preassembled to lower costs and improve quality control.

The third and final phase of the program lasted from 1995 to 1998, and witnessed a shift to extension efforts only, with the government leaving development and dissemination mostly to private actors with some oversight from local governments. The national government also began transitioning to other efforts such as providing energy efficiency labels for appliances and promoting wind turbines and solar panels. Still, by the close of the program in 1998, nearly 27,000 Ministry of Agriculture staff were engaged in rural energy promotion at the county and township levels, and there were nearly as many county level rural government agencies (2,024) as actual counties (2,126). That same year the Ministry of Agriculture claimed that more than one billion people living in 185 million of China's 236 million households had received improved stoves.

As a result, the NISP prompted a measurable decrease in the rate and extent of deforestation occurring in rural China. While the statistics are a bit dated, about four percent of China's standing forests are used as fuelwood and roughly 13 percent of cultivated land in China is used to grow fuelwood (Chan 2000). Improved cook stoves, by doubling and sometimes tripling the energy efficiency of cooking, therefore need half to one-third the amount of fuel. When combined with the results from promoting biogas stoves and diffuse firewood plantations along with improved cook stoves, Chinese efforts protected about three million hectares of young forests from being harvested each year and increased the forest coverage around rural homes about 13 percent from 1989 to 1999.

The NISP also empowered women, children, and the rural poor that would otherwise spend hours each day collecting fuelwood and cooking. Many of these individuals had to walk thousands of kilometers carrying large burdens of fuel each year, leaving less time to do other more productive and enjoyable activities and increasing the risk of health problems caused by broken bones, arthritis, exhaustion, and malnutrition. One assessment calculated that improved cook stoves in China have eliminated four million working days per year from collecting fuel spread across 1.4 million households (Catania 1999).

The Ingredients of Success

Following the "grounded" theory approach, this section elaborates on the presence of common variables across the case studies that seem to imply success at improving energy security.

Equity

Equity, or the equal distribution of benefits and/or sharing of costs, seems to play an elemental role in improving energy security. Denmark, for example, funded their research on wind energy and cogeneration through taxes so that its costs were spread among all electricity customers. Such taxes were equitable because they were based on the amount of energy consumed and funded just by ratepayers instead of all taxpayers (or externalized to society). Denmark also attempted to partially account for the "external" costs of climate change by inducing a carbon tax. Their carbon tax ensured that some of the costs related to carbon dioxide emissions were borne by those responsible instead of shifted to Danish society at large or other countries. Similarly, the Brazilian ethanol program funded itself from a tax on petroleum therefore born by all drivers. When the government mandated in 1980 that ethanol be sold at a rate cheaper than gasoline, the market difference between the two fuels was paid for by consumers themselves rather than taxpayers or the government. Grameen Shakti and the NISP in China spread program costs among consumers and users so that no single group had to pay an unfair or asymmetrical amount.

Inclusivity

Inclusion, especially the inclusion of rural actors, homeowners, and stakeholders (instead of just urban actors, commercial firms, and select groups) can enhance energy security. Inclusivity can be encouraged by the elimination of barriers to entry or access into a given sector. In Denmark, guaranteed access to the electricity grid minimized barriers to market entry and prevented utilities from using their market power or power of incumbency to block renewable energy projects (and other independently owned projects) on the grounds of inadequate transmission and distribution.

Brazil encouraged civil society groups, automobile manufacturers, oil distributors, university researchers, and many other groups to participate in the ethanol program through formal collaborations and grants, and GS shows how inclusion of stakeholders and businesses can result in effective partnerships and networks. As one example, GS has linked the use of biogas units in homes and shops with the livestock, poultry, agriculture, and fishery industries. Clients wishing to own their own biogas unit can also receive livestock, and clients that do not wish to use the fertilizer created as a byproduct from biogas units can sell it to local farmers, aquaculturists, and poultry ranchers. Similar linkages have been made in the promotion of GS's solar panels and mobile telephones, compact fluorescent light bulbs, and light emitting diode devices.

Instead of providing urban areas in China with huge chunks of energy through centralized power plants, the NISP demonstrates that relatively small changes at the rural household and village level can quickly add up and accumulate over time. Energy policy all too often focuses on influencing urban patterns of consumption due to their high population density and the relative ease of affecting large numbers of people with a single policy. Yet the NISP reminds us that much of the world's energy is consumed in rural areas where population density is low and resources are dispersed. In these areas, low-level technologies done at a small scale with a short-term investment period can be just as important as capital-intensive technologies done at a larger scale.

Information and Monitoring

Monitoring, information, and feedback also appear to play instrumental roles, at least for Brazil, Grameen Shakti, and China. The Brazilian ethanol program relied on continuous monitoring so that it could adapt to external events. Brazilian regulators set an ever-changing mandate of blending ethanol with gasoline that grew gradually to give distributors and users time to adjust. The percentages increased slowly over time (from 4.5 percent in 1977 to 15 percent in 1980 to 20 percent in 1981 and 24 percent in 1988), and also enabled targets to be raised or lowered based on market conditions. Similarly, when sales of ethanol-only vehicles started to plummet and consumers began purchasing gasoline fueled vehicles to take advantage of falling oil prices, the government responded by encouraging vehicles that could run on both ethanol and gasoline.

Similarly, GS demonstrates the necessity of monitoring and evaluation. The organization holds monthly meetings with all regional and division managers at its head office in Dhaka to discuss challenges and strategize plans for expansion. GS staff visit SHS and biogas clients an average of once a month to check on performance and proactively address potential problems. The GTCs provide quality control over SHS and biogas devices and are a source of locally produced (and higher quality) solar components. The GTCs operated by GS often conduct large demonstrations of solar and biogas devices and GS employees sometimes embark on door to door visits to familiarize communities with technology and learn about their preferences. GS engineers consistently work with village leaders to distribute brochures, hold science fairs at local elementary schools, and host workshops for policymakers.

One lesson from the Chinese NISP is the strength of continual but independent monitoring and evaluation. Each year of the NISP pilot counties were selected based on need and only 100 new counties (those with the most acute fuel shortages, requisite managerial and financial resources, and/or plentiful raw materials) were chosen per year. Cook stove designs were consistently upgraded and tested through national cook stove competitions and product certification and standardization. Officials conducted onsite inspections and checks to monitor progress and ensure villagers were using cook stoves properly.

Policy Intervention and Incentives

Each of the four case studies created conditions that incentivized energy producers and users. Danish electricity policy highlights the importance incentivizing communities to own their own wind or cogeneration plants. In 2005, for example, only 12 percent of wind farms were utility owned; the remaining 88 percent were owned by individuals and cooperatives (Toke 2005). The result is that communities are more likely to support energy systems that they have a stake in. Brazil, also, scaled back its subsidies for ethanol in order to make producers and users more cost effective. Although Brazilian regulators relied on a mix of incentives such as direct subsidies, tax breaks, preferential low-interest loans, and mandates to create a well defined and stable market at the beginning of the program, these efforts were never intended to cover all or even most of the costs associated with ethanol production. When the US ethanol industry filed a formal complaint with the US International Trade Court alleging that Brazil's subsidies for ethanol constituted dumping and illegitimate trade practices in 1985, their own independent investigation found that Brazil's subsidies amounted to less than 3 percent of total value of ethanol production.

In Bangladesh, GS relies on an installment based credit system where the users must pay for their own technology. At the beginning of their SHS, they used to require a 50 percent down payment and installments repaid in six months, but as GS became more mature they reduced payment to 25 percent down payment repaid in two years and later to 10 percent repaid in four years. GS highlights the importance of having local communities pay for renewable energy projects themselves – meaning they become invested in how they perform – but using financing to overcome the "first cost" issue through low interest loans and installments.

Analogously, in the Chinese NISP, subsidies were carefully calculated so that a majority of expenditures came from households themselves, and networks for commercial distribution were established based on market demand rather than government largesse. The national government was responsible only for certification and monitoring, with the remaining aspects of cook stove design, installation, and commercialization left to the marketplace. When overall government expenditures on the NISP are put in context, they show that the national government supported 860 counties with a total budget of only $1.4 million per year and local governments contributed only an additional $17.9 million per year. This amount corresponds to less than $25,000 spent per village. Households paid for most of the materials and installation and over the entire course of the program total government expenditure covered only about $1.28 per stove, less than one tenth the total cost of $13 per stove. The philosophy behind this component of the NISP was that people will care more about systems they have to work for and are involved with and will take more pride and responsibility in cook stove performance if they have more of their own money at stake.

Organizational Multiplicity

Each case demonstrates some of the benefits of organizational multiplicity, or having multiple regulators with overlapping jurisdiction. Danish electricity policy worked with *kommunes* when setting guidelines for cogeneration plants and wind farms, and Brazil's ethanol program received contributions from both the central government and state governments, and at times local and municipal governments. GS coordinates its activities at all scales of government in Bangladesh, enrolling the national government, state, and local governments. The NISP relied upon a network of service and implementing organizations spread across 38 provinces and municipal regions with more than 1,500 regional energy offices in 2,300 counties. Rather than fully directing the program from the top-down, responsibility was quickly shared and overlapping. The national government provided about 2.7 percent of the funds over the course of the program, provincial and local governments 13.7 percent, households and businesses 83.9 percent.

Adaptability and Flexibility

Each case study has elements of adaptability and flexibility. Denmark moderated its subsidies and support for wind and energy efficiency, scaling down and eventually repealing tax credits for wind energy in the 1990s in response to the threat of excess wind capacity. The *Proálcool* program was versatile in its production requirements. The government did not impose technology requirements on producers, and instead allowed firms to select their own production processes to meet targets. Some built new facilities from the ground up; others simply modified existing plants and retrofitted. Producers had the option of switching back and forth between ethanol and sugar production depending on market prices. If the price of sugar rose in comparison to ethanol, many plants would temporarily switch to the more lucrative opportunity and vice versa. This type of flexibility married the ethanol program to the sugar industry, and also ensured that the risks associated with producing ethanol were relatively short-lived, cheap, and reversible. GS lowered the interest they charged on their microloans when they received feedback from consumers about the way they calculated energy investments.

The NISP gave counties the authority to implement components of the program themselves and retained only the role of monitoring and evaluation for the national government. County officials focused on town-scale production and often required local technicians and manufacturers to participate, sometimes through positive incentives (such as bonuses or subsidies) and sometimes through sanctions (such as fines for producing outdated stoves or restrictions on what types of stoves could be installed in new buildings). Program officials sought feedback from users as well as local artisans involved in stove design to ensure that stoves were built to match their preferences instead of predetermined in a laboratory and imposed upon them. Stoves therefore utilized locally available materials to keep costs low rather than relying on special or imported materials; they were easily mass produced rather than individually built; and they burned wood of any size and at times additional fuels such as agricultural residue, wheat, maize, and rice stalks rather than burning only small pieces of wood.

Conclusion

In each of the cases above – Denmark's electricity policy, Brazil's national ethanol program, Grameen Shakti, and improved cook stoves in China – one can distill a common set of factors that may offer useful lessons for those wishing to respond to improve energy security. These include:

- Equity, or proportional equivalence of costs and benefits that avoid having some users get all of the benefits at little cost;
- Inclusivity, or the active involvement of a diverse number of stakeholders, usually brought about by participatory modes of decision-making or incorporation of local actors in the planning process;
- Information, which includes the distribution of data about costs and benefits, the close monitoring of programs, and mechanisms that encourage feedback and the sharing of knowledge;
- Incentives, meaning users and producers are incentivized to promote technologies that improve energy security, and also that graduated sanctions exist to enforce rules and penalize noncompliance;
- Organizational multiplicity, meaning that multiple actors are involved to deepen coverage and offer parallel systems of governance;
- Adaptability, meaning that mechanisms are in place for conflict resolution and to respond to challenges and unforeseen events.

The enduring lesson from these cases is that policymakers often focus on responding to energy problems by creating the right technology, but the study here affirms that engendering the right sort of political environment can be just as meaningful.

Appendix A Research Methods

By *case studies* the author endeavored to provide what methodological theorists Alexander George and Andrew Bennet (2004: 5) call a "detailed examination of an aspect of a historical episode to develop or test historical explanations that may be generalizable to other events." Rather than utilizing laboratory samples or statistical analysis to examine variables, case study methods involve an in-depth, longitudinal assessment of a single instance or group of instances: a case or cases.

Being *grounded* means the author commenced his research without preformed hypotheses. This method is sometimes called "grounded theory" because it is an inductive discovery method that starts with no theoretical preconception. Instead, researchers develop a conceptual account from the "ground up," the analysis grounded in the data collected itself (Strauss 1987; Glaser and Strauss 1967). A grounded approach works exceptionally well when few relevant theories yet exist to explain what is being studied, as was the situation with energy security and the four case studies.

By *ethnographic*, the author took what participants and experts said at face value, he did not prompt them, suggest answers, or intentionally inject his own values. Though ethnographic techniques are typically employed to study foreign cultures (Atkinson 1988), and intended to minimize researcher bias, the author found his own disciplinary training made him see energy issues differently than most of those he was interviewing.

For *semi-structured interviews* the author relied on a purposive sampling strategy, which means experts were chosen to represent different aspects of the cases in question, with some having expertise in public policy but others knowledge of engineering and still others, training in economics or even medicine. The author relied on semi-structured questions, sometimes referred to as "intensive interviewing" or "responsive interviewing," which asks participants a set of standard inquiries but then allows the conversation to build and deviate to explore new directions and areas (O'Sullivan et al. 2010). Many of the variables of interest to us, such as what results in improved energy security, are difficult to measure, and to describe them with quantitative methods would amount to "conceptual stretching" (King et al. 1994). Additionally, semi-structured interviews enabled the author to use face-to-face interaction, or words and images, rather than text to solicit information. The visual element of the interviews enabled the author to look for nonverbal cues to decide whether a respondent understood a question.

These methods do have some shortcomings. The qualitative aspect of interview responses makes them difficult to code and answers vary for each participant. Some respondents may provide socially desirable responses, telling the author what they think he wants to hear. Others could deliberately give us answers that they thought would sway the outcome of the study in their favor. Our sample of experts was purposive and therefore relatively small.

Appendix B List of Experts Contacted and Interviewed for the Case Studies

Case Study	Name	Institution	Location	Date
Chinese Cook stoves	Kirk R. Smith	University of California-Berkeley	Berkeley, California	June 30, 2009
Chinese Cook stoves	Peter Catania	International Energy Foundation	Okotoks, Saskatchewan	July 6, 2009
Chinese Cook stoves	Junfeng (Jim) Zhang	School of Public Health at Rutgers University	Piscataway, New Jersey	July 16, 2009
Chinese Cook stoves	Bina Agarwal	Institute of Economic Growth at Delhi University	Delhi, India	July 17, 2009
Chinese Cook stoves	Melissa Chan	Kennedy School of Government, Harvard University	Cambridge, Massachusetts	July 21, 2009
Danish Wind Energy	Steffen Nielsen	Danish Energy Authority	Copenhagen, Denmark	October 29, 2007
Danish Wind Energy	Ole Oddgaard	Danish Energy Authority	Copenhagen, Denmark	October 29, 2007
Danish Wind Energy	Anders Hasselager	Danish Energy Authority	Copenhagen, Denmark	October 29, 2007
Danish Wind Energy	Lene Nielsen	Danish Energy Authority	Copenhagen, Denmark	October 29, 2007
Danish Wind Energy	Simon B. Leonhard	Orbicon	Copenhagen, Denmark	October 29, 2007
Danish Wind Energy	Hans Henrik Lindboe	EA Energlananalyse	Copenhagen, Denmark	October 29, 2007
Danish Wind Energy	Charlotte Boesen	Dong Energy	Copenhagen, Denmark	October 29, 2007
Danish Wind Energy	Peter Wenzel Kruse	Vestas	Randers, Denmark	October 30, 2007
Danish Wind Energy	Bo Morup	Vestas	Randers, Denmark	October 30, 2007
Danish Wind Energy	Martin Molzen	LM Glasfiber	Kolding, Denmark	October 30, 2007
Danish Wind Energy	Henrik Carstens	Ramboll Denmark A/S	Virum, Denmark	October 31, 2007
Danish Wind Energy	Soren Juel Petersen	Ramboll Denmark A/S	Virum, Denmark	October 31, 2007
Danish Wind Energy	Soren A. Nielsen	MBD Offshore Power A/S	Virum, Denmark	October 31, 2007
Danish Wind Energy	Tove Feld	Det Norske Veritas	Virum, Denmark	October 31, 2007
Danish Wind Energy	Claus Gormsen	Niras	Virum, Denmark	October 31, 2007
Danish Wind Energy	Per Volund	Grontmij / Carl Bro	Virum, Denmark	October 31, 2007
Danish Wind Energy	Jesper Tornbjerg	El & Energi	Copenhagen, Denmark	October 31, 2007
Danish Wind Energy	Lise Nielson	EnergiNet DK	Copenhagen, Denmark	November 1, 2007
Danish Wind Energy	Martin Randrup	EnergiNet DK	Copenhagen, Denmark	November 1, 2007
Danish Wind Energy	Henrik Bindner	Risø National Laboratory	Roskilde, Denmark	November 1, 2007

Appendix B Continued

Case Study	Name	Institution	Location	Date
Danish Wind Energy	Sten Frandsen	Risø National Laboratory	Roskilde, Denmark	November 1, 2007
Danish Wind Energy	Jeppe Johansen	Risø National Laboratory	Roskilde, Denmark	November 1, 2007
Danish Wind Energy	Jyoti Painuly	Risø National Laboratory	Copenhagen, Denmark	November 2, 2007
Brazil's Ethanol Program	Carolina Costa	Brazilian Sugarcane Industry Association (ÚNICA)	São Paulo, Brazil	April 27, 2009
Brazil's Ethanol Program	Regis Lima Verde Leal	University of Campinas	Campinas, Brazil	May 25, 2009
Brazil's Ethanol Program	Luis Cortez	University of Campinas	Campinas, Brazil	May 25, 2009
Brazil's Ethanol Program	Arnaldo César Walter	University of Campinas	Campinas, Brazil	May 25, 2009
Brazil's Ethanol Program	Rogerio Cezar de Cerqueira Leite	University of Campinas	Campinas, Brazil	May 25, 2009
Brazil's Ethanol Program	Antonio M.F.L.J. Bonomi	Centro de Ciência e Tecnologia do Bioetanol (CTBE)	Campinas, Brazil	May 25, 2009
Brazil's Ethanol Program	Jose Pradella	Centro de Ciência e Tecnologia do Bioetanol (CTBE)	Campinas, Brazil	May 25, 2009
Brazil's Ethanol Program	Jose Goldemberg	University of São Paulo	São Paulo, Brazil	May 26, 2009
Brazil's Ethanol Program	Marcelo Galdos	University of São Paulo	São Paulo, Brazil	May 27, 2009
Brazil's Ethanol Program	Jaime Finguerut	Sugarcane Technology Center (CTC)	São Paulo, Brazil	June 2, 2009
Brazil's Ethanol Program	Patricia Simoes	Sugarcane Technology Center (CTC)	São Paulo, Brazil	June 2, 2009
Brazil's Ethanol Program	Jorge Neves	Sugarcane Technology Center (CTC)	São Paulo, Brazil	June 2, 2009
Grameen Shakti	A.K.M. Sadrul Islam	Bangladesh University of Engineering & Technology	Dhaka, Bangladesh	June 30, 2009
Grameen Shakti	Fazley Rabbi	Grameen Shakti	Dhaka, Bangladesh	July 2, 2009
Grameen Shakti	Dipal C. Barua	Grameen Shakti	Dhaka, Bangladesh	July 8, 2009
Grameen Shakti	Noim Uddin	Macquarie University	Sydney, Australia	July 9, 2009

References

Archer, D. 2009. *The Long Thaw.* Princeton: Princeton University Press.

Atkinson, P. 1988. "Ethnomethodology: A Critical Review," *Annual Review of Sociology*, 14, 441–65.

Beck, F. and Martinot, E. 2004. "Renewable Energy Policies and Barriers," in C. Cleveland (ed.), *Encyclopedia of Energy*. London and Amsterdam: Academic Press/Elsevier Science.

Bielecki, J. 2002. Energy Security: Is the Wolf at the Door? *Quarterly Review of Economics and Finance*, 42(2), 235–50.

Brown, M.A., Chandler, S., Lapsa, M.V. and Sovacool, B.K. 2008. *Carbon Lock-In: Barriers to the Deployment of Climate Change Mitigation Technologies*. Oak Ridge: National Laboratory, Oak Ridge, TN, November, ORNL/TM-2007/124.

Brown, M.A. and Sovacool, B.K. 2011. *Climate Change and Global Energy Security: Technology and Policy Options.* Cambridge, MA: MIT Press.

Catania, P. 1999. "China's Rural Energy System and Management," *Applied Energy*, 64, 238 and 240.

Chan, M. 2000. *Air Pollution from Cookstoves: Energy Alternatives and Policy in Rural China.* Pittsburg, PA: Carnegie Mellon University, 10.

Childs, B. and Bradley, R. 2007. *Plants at the Pump: Biofuels, Climate Change, and Sustainability.* Washington, DC: World Resources Institute.

Committee on Climate Change Science and Technology Integration (CCCSTI) 2009. *Strategies for the Commercialization and Deployment of Greenhouse Gas-Intensity Reducing Technologies and Practices*. Washington, DC: US Department of Energy.

Cordonnier, V.M. 2008. "Ethanol's Roots: How Brazilian Legislation Created the International Ethanol Boom," *Environmental Law & Policy Review*, 33(1), 287–318.

Dietz, T., Ostrom, E. and Stern, P. 2003. "The Struggle to Govern the Commons," *Science*, 302(5652), 1907–12.

Florini, A.E. and B.K. Sovacool. 2009. "Who Governs Energy? The Challenges Facing Global Energy Governance," *Energy Policy* 37(12) (December, 2009), 5239–48.

Folke, C., Hahn, T., Olsson, P. and Norberg, J. 2005. "Adaptive Governance of Socio-Ecological Systems," *Annual Review of Environment and Resources*, 30, 441–73.

George, A.L. and Bennett, A. 2004. *Case Studies and Theory Development in the Social Sciences.* Cambridge, MA: Harvard University Press.

Glaser, B.G. and Strauss, A.L. 1967. *The Discovery of Grounded Theory.* Chicago, IL: Aldine.

O'Sullivan, E., Rassel, G.R. and Berner, M. 2010. *Research Methods for Public Administrators.* New York: Pearson Longman.

Goldemberg, J. 2006. "The Ethanol Program in Brazil," *Environmental Research Letters*, 1(1), 1–5.

Goldemberg, J. 2008. "The Brazilian Biofuels Industry," *Biotechnology for Biofuels*, 1(6), 1–7.

Hardin, G. 1968. "The Tragedy of the Commons," *Science*, 162(3859), 1243–7.

Hardin, R. 1982. *Collective Action.* Baltimore, MD: Johns Hopkins University Press.

Hooghe, L. and Marks, G. 2003. "Unraveling the Central State, but How? Types of Multi-Level Governance," *American Political Science Review*, 97(2), 233–43.

International Energy Agency. 2010. *Key World Energy Statistics.* Paris: OECD.

Islam, A.K.M., Sadrul, M.I., and Rahman, T. 2006. "Effective Renewable Energy Activities in Bangladesh," *Renewable Energy*, 31(5), 677–88.

Islam, K.R. and Weil, R.R. 2000. "Land Use Effects on Soil Quality in a Tropical Forest Ecosystem of Bangladesh," *Agriculture, Ecosystems and Environment*, 79(1), 9–16.

King, G., Keohane, R.O. and Verba, S. 1994. *Designing Social Inquiry: Scientific Inference in Qualitative Research.* Princeton, NJ: Princeton University Press.

Leite, R.C.D.C., Manoel R.L.V.L., Luis A.B.C., Griffin, W.M., and Mirna, I.G.S. 2009. "Can Brazil Replace 5% of the 2025 Gasoline World Demand with Ethanol," *Energy*, 34(5), 655–61.

Lijphart, A. 2004. "Constitutional Design for Divided Societies," *Journal of Democracy*, 15(2), 96–109.

McGinnis, M.D. 2005. "Beyond Individualism and Spontaneity," *Journal of Economic Behavior and Organization*, 57(2), 167–72.

McGinnis, M.D. 2006. *Costs and Challenges of Polycentric Governance.* Revised version of paper presented at the Workshop on Analyzing Problems of Polycentric Governance in the Growing EU, Humboldt University, Berlin.

Miah, D., Al Rashid, H. and Man Yong Shin 2009. "Wood Fuel Use in the Traditional Cooking Stoves in the Rural Floodplain Areas of Bangladesh: A Socio-Environmental Perspective," *Biomass and Bioenergy*, 33(4), 70–78.

Morthorst, P.E. 2000. "The Development of a Green Certificate Market," *Energy Policy*, 28(15), 1085–94

Olson, M. 1965. *The Logic of Collective Action.* Cambridge, MA: Harvard University Press.

Painuly, J.P. 2001 "Barriers to Renewable Energy Penetration, a Framework for Analysis," *Renewable Energy*, 24(1), 73–89.

Poteete, A., Janssen, M. and Ostrom, E. 2010. *Working Together: Collective Action, the Commons, and Multile Methods in Practice.* Trenton, NJ: Princeton University Press.

Qiu, D., Shuhua G., Catania P., and Huang K. 1996. "Diffusion of Improved Biomass Stoves in China," *Energy Policy*, 24(5), 463–9.

Schumacher, E.F. 1977/ *Shumacher on Energy: Speeches and Writings of E.F. Schumacher*. London: Cape.

Sinton, J.E., Smith, K.R., Peabody, J.W., Yaping, L., Xiliang, Z., Edwards, R., and Quan, G. 2004. "An Assessment of Programs to Promote Improved Household Stoves in China," *Energy for Sustainable Development*, 8(3), 33–52.

Smith, K.R. Shuhua, G., Kun, H., and Daxiong, Q. 1993. "One Hundred Million Improved Cookstoves in China: How Was it Done?" *World Development*, 21(6), 941–61.

Sovacool, B.K. 2008a. *The Dirty Energy Dilemma: What's Blocking Clean Power in the United States*. Westport, CT: Praeger.

Sovacool, B.K. 2008b. "The Best of Both Worlds: Environmental Federalism and the Need for Federal Action on Renewable Energy and Climate Change," *Stanford Environmental Law Journal*, 27(2) (June), 397–476.

Sovacool, B.K. (ed.). 2010. *The Routledge Handbook of Energy Security.* London: Routeldge.

Strauss, A.L. 1987. *Qualitative Analysis for Social Scientists.* Cambridge: Cambridge University Press.

Toke, D. 2005. "Are Green Electricity Certificates the Way Forward for Renewable Energy? An Evaluation of the United Kingdom's Renewables Obligation in the Context of International Comparisons," *Environment and Planning C*, 23, 361–74.

United Nations Development Program. 2010. *Energy for a Sustainable Future: The Secretary-General's Advisory Group on Energy and Climate Change Summary Report and Recommendations.* New York: UNDP, April 28.

Chapter 4
Global Energy Governance Post-2012: Bridging the Gap Between Energy Security and Climate Policy

Meraiah Foley

Introduction

Energy security and climate policy are frequently presented as two sides of the same issue. While these two policy areas share some common goals, they are as likely as not to clash. A wholesale transition from fossil fuels to renewable resources would certainly reduce the risks associated with both climate change and long-term energy security, but the deployment of capital needed to make this transition is immense and will take decades. What happens in the meantime has profound implications for both energy security and climate policy.

The first half of this chapter will argue for the need to find a bridge between energy security and climate policy. The wrong policy mix – too far weighted in favour of either objective – could have disastrous consequences for the economy, geopolitical stability and the environment. To whatever degree possible, lawmakers must exploit solutions where the goals of energy security and climate policy converge. Although it is not a silver bullet, there is broad consensus among both energy and climate experts that energy efficiency represents the least expensive, least controversial approach to slowing energy demand. But there has, as yet, been no serious, concerted effort to improve energy efficiency on a global scale. Within the climate policy mix, the primary tool for incentivising the uptake of low-carbon technology – the Clean Development Mechanism (CDM) – has largely failed to capitalise on the emissions saving potential of energy efficiency.

The second half of this chapter will examine the CDM in terms of its potential for increasing energy efficiency in the developing world, where energy consumption is growing fastest. The section begins with an overview of the mechanism and the criticisms of its application to date. Next, this chapter will examine the CDM in the context of energy efficiency, and discuss how the mechanism should be reformed in the post-2012 framework. The objective of this chapter is to argue that energy efficiency represents a relatively inexpensive, uncontroversial area of convergence between energy security and climate policy; and to examine the ways in which one existing multilateral instrument – the CDM – could be better employed to pick this "low hanging fruit."

Strange Bedfellows: The Tension between Energy Security and Climate Policy

To win support for climate policies that have the potential to send energy costs soaring, politicians frequently invoke the goal of energy independence – defined as a lack of reliance on imported fuel – arguing that energy security and climate change are "two sides of the same coin."[1] In their rhetoric, these leaders often conflate the goals of energy security, energy independence and climate change mitigation in ways that are not always practical, logical or true. While there is overlap in some areas of climate and energy policy, in reality – for many energy-importing countries – the objectives are as likely to clash as they are to converge.

Energy security is typically preoccupied with maintaining a country's access to affordable, reliable and adequate energy supplies (IEA 2009), though the concept of "energy security" has different meanings for different stakeholders (Chester 2010; von Hippel et al. 2010). For energy-importing states – that is, most of the world's major economies – energy security policy aims largely to hedge against short-term supply disruptions (caused by natural disasters or politically-motivated strikes), and to limit dependence on potentially hostile or unstable suppliers. These concerns are all underpinned by a mountain of evidence linking the affordability and accessibility of energy – particularly oil and gas – to a country's economic and military security.[2] While the idea of "environmental acceptability" (Kruyt et al. 2009) is gaining traction as an indicator of long-term energy security, the most commonly accepted measures of energy security are: import dependence (particularly from unstable regions); the distance from production to consumption; the vulnerability of supply infrastructure to disruption from natural disasters or political attacks; the degree of fuel substitutability; the diversity of the fuel mix; and the concentration of market power among energy providers (IEA 2009: 115).

Climate policy, on the other hand, has been dominated by the push to reduce emissions of dangerous greenhouse gases (GHGs) to prevent potentially catastrophic global temperatures. The objectives of climate policy are supported by volumes of scientific research indicating that, without major changes to the way humans consume energy, GHG emissions will rise by 40 percent between 2007 and 2030, putting the world on a path to a potentially catastrophic temperature increase of 6 degrees Celsius (IPCC 2007; IEA 2009). While there is some disagreement about what constitutes a sustainable long-term temperature rise, a general consensus has emerged that efforts should be made to limit the temperature increase to 2 degrees Celsius (IEA 2009: 198). Stabilising atmospheric concentrations of GHGs to 450 parts per million (ppm) of carbon dioxide and other gases (referred to collectively as CO2 equivalent, or CO2e) provides a 50 percent chance of keeping the global temperature at 2 degrees Celsius.[3]

With the dramatic increase in energy consumption seen from the rise of China, India and other rapidly developing economies, energy security and climate change are emerging as two of the world's most complex policy challenges – but often for different reasons. While there have been some notable attempts to identify the areas of convergence,[4] energy security policy remains largely focused on the economic, geopolitical and military implications of increased fossil fuel demand, while the scientific community is becoming increasingly alarmed by the dramatic rise in emissions.

In the idealised rhetoric of many politicians, energy security and climate change policy are "two sides of the same coin" that can be addressed by a single policy response. The reality, however, is much more complicated. While there is some convergence between climate policy and energy security, it is important for policy makers to recognise that they are also frequently at odds. Certainly, a complete switch from fossil fuels to renewable resources (like wind and solar) would greatly reduce the risk of climate change as well as the risks associated with import-dependence. But this transition, however desirable, will not occur overnight. The deployment of capital necessary to manage such a transition is immense, and will likely take decades.[5] What happens in the meantime – which is critical if the world is to keep atmospheric concentration of GHGs at 450 ppm CO2e – has profound implications for climate change mitigation and energy security policy.

From an energy security perspective, limiting the use of coal, for example, could have serious negative consequences for the many large economies that have it in abundance.[6] Increased reliance on natural gas as a cleaner alternative to coal in the production of electricity could increase the relative power of Iran and Russia, which hold about 50 percent of the world's known reserves.[7] Similarly, restricting the uptake of emissions-intensive unconventional fuels (such as shale oil and the Canadian tar sands) would increase the world's dependence on oil from the Middle East (Hartley and Medlock 2008). Other studies have shown that a global policy mix based around the 450 ppm CO2e scenario would have a detrimental impact on energy prices (Kruyt et al. 2009). But ignoring the environmental consequences of energy policy puts the entire planet at risk.

There is a small but growing body of literature examining the interaction between climate change policy and energy security (von Hippel et al. 2010; Kruyt et al. 2009; Hartley and Medlock 2008; Blyth and Lefèvre 2004). The aim of this section is not to assess the relative strengths and weaknesses of each position, but to illustrate that climate policy and energy security policy are frequently misaligned. Climate policy makers who neglect the economic and geopolitical consequences of their actions run the risk of plunging the world into recession or increasing the world's dependence on potentially hostile, unstable regimes. On the other hand, energy security extremists who believe that environmental risk should remain secondary to economic and geopolitical concerns risk the economic consequences of ignoring climate change in the medium-term, and gamble with the long-term habitability of the planet. Striking the right balance between energy security and climate policy is critical to ensure the strength of the global economy, the stability of the geopolitical order and the health of the environment.

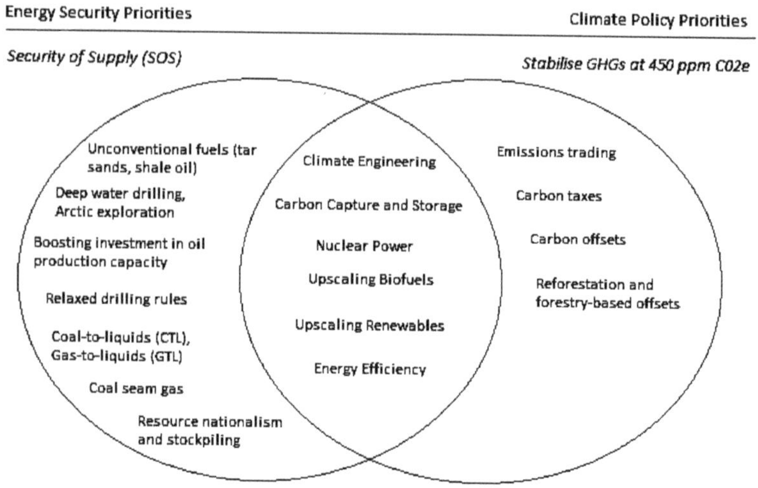

Figure 4.1 Energy Security and Climate Policy: Bridging the Gap
Source: Author.

The Promise of Energy Efficiency

Despite the conflicts between energy security and climate policy, there are many areas of convergence (see Figure 4.1). Many of the policy responses that best address both energy security and climate change have significant drawbacks, however. Carbon capture and storage (CCS) and climate engineering are as yet unproven and untested on a large scale. Nuclear power has numerous associated risks, and the rapid uptake of biofuels in the mid-2000s created significant negative externalities for global food production and the preservation of forests. Climate adaptation measures, such as reinforcing levees and protecting coastal refineries from storm surges, deal only with the medium-term disruptions likely to be wrought by climate change; they do not address the longer-term health and habitability of a significantly warmer planet.

The most promising area of convergence is energy efficiency, the process of using less energy to produce the same amount of goods or work. Scientists at the Intergovernmental Panel on Climate Change (IPCC), have concluded, with a high level of agreement, that energy efficiency measures could reduce global emissions by 29 percent of 1990 levels by 2020 in the residential and commercial sectors, *the highest return among all of the sectors studied* (IPCC 2007: 389, emphasis added). Similarly, the International Energy Agency (IEA) has issued 25 policy recommendations which, if implemented, could reduce global emissions by 20 percent per year, or 8.2 Gt CO2/year, from 2005 to 2030 (See also: Gupta and

Ivanova 2009; Arquit Niederberger 2008; IEA 2008; Expert Group on Energy Efficiency 2007; Hinostroza et al. 2007) .

In the building sector – which accounts for about 40 percent of energy use is most countries – governments could reap emissions reductions of up to 1.4 Gt CO_2/year by 2030 simply by altering building codes to make new structures more energy efficient; actively promoting the uptake of passive energy housing and zero energy buildings; and setting higher efficiency standards for the renovation of existing stock (Ibid). The transportation sector – which accounts for about 60 percent of world oil consumption – could also benefit from advances in energy efficiency. By setting mandatory fuel efficiency standards for light and heavy-duty vehicles, legislating rolling resistance standards for tyres and promoting appropriate tyre inflation pressure guidelines, the IEA projects that governments could reduce fuel consumption and resulting emissions by 1.4 Gt CO_2/year by 2030 (Ibid). Electricity consumption from lighting could be reduced by 38 percent with the widespread use of more efficient lighting technologies, such as compact fluorescent lamps (CFLs), which are four to five times more efficient than conventional incandescent globes. According to the IEA, savings from such a measure could equal up to 1.2 Gt CO_2/year by 2030 (Ibid).

These savings alone are not sufficient to ensure long-term energy security, or mitigate climate change. Difficult policy choices will still have to be made. But there is widespread consensus that energy efficiency offers the biggest benefits to both policy areas, at the lowest risk (Hartley and Medlock 2008: 47; IEA 2008; IEA 2009; IPCC 2007; Blyth and Lefèvre 2004: 68–69). The degree to which such energy efficiency measures would favour a country's energy security or climate policy goals depends on its circumstances and regulatory mix. For energy importing states, efficiency measures in the transport sector would tend to have a stronger effect on energy security – by reducing the dependence on oil – while efficiency measures in the electricity sector would deliver a greater payback in emissions reduction by reducing the use of coal-fired power (Blyth and Lefèvre 2004: 68).

In addition, energy efficiency projects confer several other benefits. In both developed and developing countries, they have the potential to foment jobs creation and new business opportunities related to the retrofitting of buildings; increased economic competitiveness; social welfare benefits for low-income households via reduced electricity or fuel bills; savings for governments in developing countries that subsidise energy consumption; increased access to energy services; and reduced rates of illness and death from indoor air pollution associated with the use of inefficient cook stoves and lighting in developing countries (IPCC 2007: 389–90).

Nevertheless, global uptake of energy efficiency measures has been slow, and serious discussion of energy efficiency policy has been largely absent from the mainstream debates on climate change mitigation and energy security (Gupta and Ivanova 2009; Arquit Niederberger 2008). The adoption of energy efficiency has been hamstrung by a number of barriers. Replacing older equipment with newer, energy efficient alternatives can deliver energy cost savings, but if the capital and maintenance costs of the old equipment are significantly less, uptake of the new technology can be low. Energy efficiency measures are often small and incremental in scale, and require the participation of many actors. Without the coordinating hand of national or international policy, the correct incentives and large public information campaigns, consumers and businesses continue with the status quo (Gupta and Ivanova 2009: 340–41). Although there has been some local regulatory movement in this area, there has been an overall lack of political will to fully exploit the energy efficiency's potential in both emissions reduction and resource security.[8] A lack of vested interests promoting energy efficiency is also a factor (Gupta and Ivanova 2009).

Targeting the Poor: Why Energy Efficiency in the Developing World Matters

The IEA's reference scenario – a picture of what the future would look like if current energy patterns were to continue unabated – predicts that global energy demand will rise by 40 percent from 2007 to 2030, reaching 16.8 billion tons of oil equivalent (toe). China, India and the Middle East account for more than 80 percent of the increase. In all, 90 percent of the increase in global energy demand is expected to come from developing countries outside the Organisation for Economic Cooperation and Development (OECD) (IEA 2009: 74–6). Fossil fuel emissions in this scenario also increase by 40 percent, from 29 Gt in 2007 to 40.2 Gt in 2030, a level of emissions that would place the world on the path to a global average temperature rise of 6 degrees Celsius (IEA 2009: 113). Of the 11 Gt additional emissions under the reference scenario, China accounts for 6 Gt, India for 2 Gt and the Middle East for 1 Gt (IEA 2009: 167).

Under an alternate scenario in which energy consumption is constrained to hold atmospheric GHG concentrations at 450 ppm CO2e, primary energy demand still grows by 20 percent between 2007 and 2030, led again by China, India and the Middle East. Although, in this scenario, zero-carbon energy sources comprise 32 percent of the total primary energy mix in 2030, fossil fuels still account for the remaining 68 percent (IEA 2009: 195).

With the failure of the United Nations Conference of the Parties (COP) to reach an agreement on a successor to the 1997 Kyoto Protocol in Copenhagen, Denmark, the success of such a scenario is still in doubt. Nevertheless, there remains a pressing need to curtail emissions growth and energy consumption in developing countries – without unduly restricting their economic growth. Although developing countries are responsible for 42 percent of the cumulative GHG emissions since 1890 (IEA 2009: 6), they represent the lion's share of new emissions. In addition to their rapid growth, these economies also tend to have higher levels of energy intensity – the amount of energy needed to produce a single unit of gross domestic product (EIA 2006). Although China has an ambitious program to cut its energy intensity, inefficiency in most of Asia and the Middle East remains high[9] (Ibid; Bradsher 2010). In contrast, the 28 member states of the IEA use about 45 percent less energy to produce a unit of GDP than they did in the early 1970s, before the oil price shock of the 1973 embargo (Blyth and Lefèvre: 68). This decrease has been driven by widespread attempts to improve energy efficiency, particularly in resource-poor countries like Japan, South Korea and France, and periodic drives toward energy independence in resource-rich countries like the United States, Australia and Canada (von Hippel et al. 2009: 3).

Energy efficiency, with its emphasis on doing the same amount of work with less energy, presents an ideal policy response for the developing world, which has been strongly reluctant to join the international framework on climate change for fear that limiting fossil fuel consumption will slow economic growth. Energy efficiency confers many other benefits to poor countries: it is less expensive than creating new supply; it frees up resources for other uses, like social development; it can help reduce the amount of energy subsidies paid by governments in many developing countries to poorer consumers; and it has the potential to increase access to energy services and reduce human exposure to air pollution from inefficient wood fuels and indoor cook stoves (IPCC 2007: 59, 389–90). There is a pressing need to improve energy efficiency in both the developed and developing world. But in the context of the latter's rapid emissions growth and high levels of energy intensity, scaling up energy efficiency in the developing world is particularly important.

Global Energy Governance and Energy Efficiency

Alongside the high-profile negotiations about the future of the global climate treaty, a quieter, less publicised discussion has emerged about the need for an international framework to manage the tensions that are expected to rise among and between energy producing and energy consuming states as the supply of fossil fuels grows increasingly constrained. Most of this literature is preoccupied with the geopolitical, economic and military risks posed by the rapid increase in demand for fossil fuel resources, and the impact of China and India's resource strategies on the rest of the world (see Moran and Russell 2009; Wenger et al. 2009; Klare 2008). Despite the clear effects that a global climate framework might have on the energy security realm, there has been relatively little effort within the energy governance debate to find convergence between climate governance and long-term energy security policy.[10]

This chapter does not argue that climate policy is the sole answer to global energy security. But the overwhelming focus on the geopolitics of oil and gas by those leading the global energy governance debate ignores existing avenues for cooperation and problem solving. With its calls for a hypothetical new energy framework, the global energy governance debate has been too long separate from the material progress that has been made in climate governance. Wherever mechanisms exist that have the potential to bridge the gap between long-term energy security and climate change abatement, they should be exploited to the fullest extent possible. The next section looks at one such instrument that has not yet reached its full potential.

The Clean Development Mechanism: Carbon Finance and the Developing World

Overview of the Mechanism

The Clean Development Mechanism (CDM, or "the mechanism") was established by the 1997 Kyoto Protocol, under which the parties to the United Nations Framework Convention on Climate Change (UNFCCC) agreed that developed countries should reduce their emissions of six greenhouse gases[11] by at least 5 percent of 1990 levels between 2008 and 2012.[12] To help these industrialised countries (Annex I countries) meet their targets, three flexibility mechanisms were introduced into the scheme. The flexibility mechanisms allow Annex I countries meet their binding emissions targets by contributing to emissions-reducing projects in developing countries (Non-Annex I countries) through the CDM,[13] or in other Annex I countries where the price of emissions abatement is cheaper through the Joint Implementation, or by purchasing emissions credits (Assigned Amount Units) from other developed countries The CDM was intended to pave the way for developing countries – which do not have binding emissions reductions targets – to participate in the scheme. The idea was for developed countries with binding emissions targets to help developing countries make the transition to lower-carbon energy systems via financial investment or technology transfer, in exchange for Certified Emissions Reductions (CER) credits that could be used to offset a portion of their own emissions. Thus, the CDM was created to meet two objectives: helping Annex I countries meet their binding emissions targets and promoting sustainable growth in the developing world.[14]

The CDM is administered by an Executive Board under the auspices of the UNFCCC. The CDM has gained momentum since the Kyoto Protocol took effect in 2005. Since then, the board has registered more than 2,200 projects that are expected to produce CERs amounting to more than 2.9 billion tonnes of CO2e.[15] In 2008, US$6.5 billion flowed to offset projects through the mechanism, while secondary sales of CDM credits – the buying and reselling by brokers and investors – reached US$25 billion (Capoor and Ambrosi 2009: 1). Yet despite its popularity as a source of carbon offsets for Annex I countries, the mechanism has become one of the most maligned aspects of the Kyoto Protocol, and its future is a sticking point in negotiating a successor to the protocol, which expires at the end of 2012.

Criticisms of the CDM

The CDM was intended as a tool for financing the transfer of low-carbon technologies from the developed to the developing world, thereby assisting poorer countries to leapfrog a cleaner, less carbon-intensive phase of development. But the mechanism has largely failed to deliver this result, favouring large, one-off industrial upgrades that have delivered questionable emissions savings and made negligible contributions to sustainable development (see Lloyd and Subbaro 2009; Boyd et al. 2009; McCully 2008; Wara 2008; Figueres 2006). Critics argue that the mechanism's strict rules for monitoring and verification have created a situation in which the aim of helping Annex I countries comply with their emissions targets has superceded the sustainable development objective.

Cost is one barrier. The expense and time needed to shepherd a CDM project through the Executive Board's lengthy validation process has tended to favour large-scale, high volume projects over smaller community-based initiatives that have the potential to deliver significant emissions reductions, but which are harder to measure (Lloyd and Subbaro 2009: 240; Arquit Neiderberger 2008; Hinostroza et al. 2007; Figueres 2006).[16] Thus, most of the projects that have gone through the CDM pipeline represent isolated upgrades to a single plant or factory, rather than broad-based efforts to decarbonize the energy supply. These high costs have left many of the world's poorest countries in the cold. India, China, Brazil and Mexico have hosted 75 percent of all registered projects, while 90 percent of emissions reductions credits have been issued to China, India, South Korea and Brazil. Africa has received just 2.5 percent of CDM funding (UNEP 2010).

Environmental legitimacy was intended to be a core principle of the CDM, but several studies have argued persuasively that many of the emissions credits produced by the CDM do not represent real changes in energy use (McCully 2008; Wara 2008). These early critiques have focused on the CDM's disproportionate approval of projects dedicated to the destruction of trifluoromethane (HFC-23), a GHG associated with refrigeration that is several times more potent than CO2. Because each emissions credit is measured in tons of CO2 abated, and because HFC-23 can be removed from the industrial process with relatively minor technological upgrades, the CDM created a perverse incentive whereby manufacturers could make more money selling emissions credits than from selling their own product. This spurred enterprising investors in China to build new refrigerant plants just to cash in on the CDM windfall (Fein et al. 2010; Wara 2008). The Executive Board has taken steps to rectify this well-documented problem. But to date, 74 percent of the CER credits generated under the scheme have been from the destruction of HFCs and other gases, compared to 13 percent for renewable energy projects and less than 5 percent for energy efficiency projects (UNEP 2010).

Critics have also questioned the effectiveness of the CDM's "additionality" requirement in generating real changes to energy use. To ensure environmental legitimacy, the Kyoto Protocol stipulated that all projects must be "additional" to what would have occurred in the absence of the CDM incentive.[17] The intent of this clause was to ensure that Annex I countries could not offset their own emissions with projects that would have gone ahead anyway. Establishing additionality requires setting a baseline methodology to determine the level of emissions that would have occurred in the absence of the CDM. Baseline measurements must be project specific, and defined such that emissions credits cannot be generated for emissions reductions that occur outside the project context, such as changes in national policy (Michaelowa et al. 2009: 355). This has put broad-based energy efficiency programs at a distinct disadvantage, which will be discussed in greater detail in the next section. Critics argue that the mechanism's project-by-project approach has created results that make sense on an individual basis, but which are illogical in the aggregate (Fein et al. 2010). For example, early reviews of the scheme found that the Executive Board approved nearly every new non-coal power plant in China, though it is very likely some of those projects would have gone ahead without CDM funding (Wara 2008). Another study found that only 1 percent of CDM projects made a measurable contribution to sustainable development (Sutter and Parreno 2007: 85–7).

The CDM and Energy Efficiency

The potential of energy efficiency to reduce energy consumption and emissions growth is almost universally recognised. As discussed earlier in this chapter, energy efficiency projects are of particular interest to developing countries because – in addition to their GHG abatement potential – they have the power to contribute to economic and social objectives, as well as promoting energy security by reducing the overall demand for fossil fuels. Yet energy efficiency measures have not fared well under the CDM, due to the administrative complexities mentioned above (Gupta and Ivanova 2009; Michaelowa et al. 2009; Arquit Niederberger 2008; Carbon Trust 2008; Figueres 2006).

Energy efficiency projects in the CDM fall into two main categories: supply side projects – which tend to focus on industrial processes like aluminium production, oil refining or power generation – and demand side, or end-use energy efficiency. There have been just 189 end-use efficiency projects out of more than 5,000 projects the CDM's pipeline. By 2012, these projects (some of which are still awaiting approval) could generate a combined emissions reduction of 24.3 million tonnes of CO2e (UNEP 2010). That stands in stark contrast to the 8.2 Gt CO2e per year that could be realised by 2030 if the IEA's 25 efficiency recommendations were adopted globally. While these numbers are reflect emissions abated, it is important to also recognise that they represent real energy savings through reduced demand, and therefore have implications for energy security as well as climate policy.

Only a handful of CDM projects have been approved in the area of energy efficient lighting, though the IEA estimates that electricity consumption could be reduced by 38 percent with the widespread use of this technology. Transportation and end-use energy efficiency projects in the service and industrial sectors have also been under-represented in the CDM, with a combined total of 181 projects at registration or validation (UNEP 2010). Supply-side projects (such as recycling waste heat into electricity) have fared better, with 461 projects in the pipeline, are still under-represented relative to their potential for reducing energy demand.[18]

Efficiency projects, particularly on the demand side, have faced significant barriers under the CDM's rules for establishing baselines and additionality – measurements that affect the level of emissions reductions that can be "sold" as offsets to Annex I countries. One recent study found that projects aimed at end-use energy efficiency (such as cleaner cooking stoves or energy efficient lighting) have significantly higher rejection rates than other types of projects because tracking the emissions reductions produced by these small, widely dispersed devices is costly and impractical (Michaelowa et al. 2009: 357). Moreover, the additionality criterion creates a perverse disincentive for the sort of broad-based policy changes advocated by the IEA. Some critics have warned that developing countries that enact mandatory energy efficiency standards could miss out on CDM funding because projects that seek to upgrade their technology to meet those guidelines could be excluded under the baseline requirement (Figueres 2006: 12). Thus, a state that moves to improve energy efficiency by decree could see its efficiency-related projects fail the additionality test, without regard to the projects' overall merits. The Executive Board has attempted to address this problem by exempting national or sectoral policies issued since 2001,[19] but the additionality requirement remains unclear in practice (Arquit Niederberger 2008: 133).

These strict measurement and reporting requirements have distorted the CDM in favour of its first goal – helping Annex I countries meet their emissions targets by delivering discrete, measurable offsets – over the kind of sweeping market transformation needed to bring about real changes in energy use patterns (Arquit Niederberger 2008: 130–31). In other words, the mechanism's one-eyed focus on accuracy, precision and attribution has come at the expense of broader climate goals, particularly in the realm of energy efficiency. From the perspective of industrialised countries trying to meet their Kyoto targets, the CDM has been a success. But it has failed to address even the most basic sustainable development goals for all but a handful of developing states. Its contribution to global emissions reduction is questionable. And, though it was not designed explicitly for this purpose, it has also failed to meet its potential as a tool for enhancing energy security.

Reforming the CDM: Exploiting the Promise of Energy Efficiency

There are several promising areas for reform. With the successor to the Kyoto Protocol in doubt, the future of the CDM is also uncertain. But despite the criticisms against it, the mechanism has become one of the most widely-employed aspects of the protocol. As such, it is unlikely to disappear altogether from future climate agreements. Numerous pathways have been suggested for reforming the

CDM, two of which are discussed in light of their potential for rolling out major improvements to energy efficiency in the developing world.

Programmatic and Sectoral CDM

In response to the criticism with respect to smaller-scale projects, the Excutive Board has approved methodologies for programs of activities (PoA) under the CDM. This so-called programmatic approach (or p-CDM) allows several similar CDM projects to be bundled together to help reduce the processing costs associated with establishing baseline and additionality requirements. Thus, multiple projects that are similar in design can be registered as a single CDM initiative, provided they meet the methodological requirements (Hinostroza et al. 2007). Once approved, the project can roll out an unlimited number of activities over a 28-year period. The advent of p-CDM has been hailed as an important development for end-use energy efficiency projects because it can significantly reduce administrative costs associated with registration and monitoring (Michaelowa et al. 2009: 361–2).

There have only been a handful of p-CDM projects to date – including at least two end-use efficiency project involving the distribution of energy efficient lighting – so it is difficult to measure their success. But some early studies suggest that while the p-CDM approach can help reduce the cost barrier for smaller, dispersed end-use efficiency projects, there are still significant methodological hurdles to overcome. How to determine additionality in energy efficiency projects, monitor many dispersed appliances (such as light globes), and how to differentiate exogenous factors from the effect of the project are questions that are still largely unanswered in this area.

In light of the CDM's failure to promote sustainable development, others have advocated the adoption of a sectoral approach whereby developing countries would be encouraged to establish regional, sectoral, sub-sectoral or cross-sectoral project activities which would stem from "specific sustainable development policies" (Figueres 2006: 14 citing Samaniego and Figueres 2002). This approach, it was argued, would encourage top-down implementation of sustainable practices in response to changes in government policy. Unlike the current CDM, in which emissions reductions are measured by individual projects, a sectoral CDM (or s-CDM) sets the *policy as the project* (Ibid: 17) and CERs are only generated where a sector's emissions fall below an established baseline. Under such a scheme, a regional push to improve fuel efficiency in industrial motors or set tighter efficiency guidelines for new buildings via a new legal standard could qualify under the s-CDM. Unlike in the current CDM, each individual emissions-reducing activity (e.g. building retrofit or new motor) contributes to the overall emissions reduction, but does not, by itself, constitute a measurable project.

The s-CDM approach would help address the problem of determining additionality in end-use efficiency projects by setting a list of policies that would *de facto* be additional. Such a list might include, for example, policies that set energy efficiency standards, or provide incentives or subsidies to encourage the uptake of energy efficiency measures in buildings, vehicles, lighting and appliances. Such measures would be considered additional "by virtue of their decided GHG reduction effect, and the fact that these activities usually require a supporting policy in order to occur on a significant scale" (Figueres 2006: 22). This would eliminate the cumbersome need to monitor each individual efficiency unit (e.g. light globe or vehicle), as the aggregated result would be deemed additional because it would not have occurred in the absence of the policy change.

Both the programmatic and sectoral approaches to the CDM could contribute to the proliferation of energy efficiency measures. But regardless of whether the future CDM takes a programmatic, sectoral or policy approach, the mechanism's baseline and additionality criteria need to be substantially altered. The environmental benefit or integrity of any project or PoA must be assessed against the broader context to prevent projects of questionable merit (such as the HFC-23 upgrades) receiving favour over energy efficiency measures that are harder to quantify but have the potential to deliver greater returns at less expense. The CDM of the future must recognise that simpler, but still robust and conservative methodological approaches to the quantification of energy savings can result in durable benefits to the mitigation of climate change (Arquit Niederberger 2008: 136). The CDM has the potential to deliver substantial results in the area of energy efficiency, but it must be complemented by extensive policy reforms in the transport, industry and building sectors. A reformed mechanism that can incentivise these policy reforms will have the greatest chance of fomenting the broad transformation in energy use patterns needed to mitigate climate change and ensure long-term energy security.

A Clean Development Fund

Another possibility is to replace the current CDM with a Clean Development Fund (Fein et al. 2010; Wara 2008). Proponents of this view argue that the CDM's mechanism model exacerbates its administrative inefficiencies, and relegates the Executive Board to a passive role by limiting its discretion to a simple "yes" or "no" analysis of any project that purports to meet its narrowly defined criteria. A fund-based system would greatly reduce the number of players and steps in the system, as financing would flow directly from the fund to the project and back again. This system would also recast the Executive Board as active fund managers who would have the luxury of financing only those projects that are most environmentally sound, rather than approving every project that claims to meet the CDM's criteria (Fein et al. 2010: 9). In making their yearly funding decisions, the managers could look at all of the project proposals together and then pick the most promising ones. This more holistic approach would ostensibly make it easier to spot efforts to game the system (as with the HFC-23 upgrades). However, advocates of this approach acknowledge that it would not preclude fund managers from funnelling money to bad projects, whether through innocent ignorance or wilful corruption (Ibid: 10), though these problems could be alleviated with strict oversight and protective design.

One such measure could be to allocate funds via a reverse auction (Wara 2008), whereby projects would be assigned priority according to their rank on a cost-per-abatement scale. Such a process would ensure that funding would flow only to projects that deliver the biggest return on the dollar. However, winning the reverse auction would not automatically confer a right to funding, but would allow fund managers to compare projects on a cost per emissions abatement basis and apply a more comprehensive approach to mitigation efforts during any given funding cycle.

A discretionary, fund-based system has the potential to improve the uptake of energy efficient projects. By giving fund managers more discretionary power, the fund-based approach would make it easier to underwrite projects with the greatest energy saving potential. The use of reverse auction would temper this discretion by assigning priority weight to projects that are deemed to have the biggest energy savings potential at the lowest cost. For example, the energy efficiency measures recommended by the IEA, which have so far remained largely outside the context of the CDM, could receive priority weight according to their universally recognised emissions savings potential. This would send a clear signal to investors and applicants to the fund, offering the widespread incentives that have so far been lacking in the energy efficiency sphere. Fund managers would have the discretion to change these weighting measures, with input from the public. This could potentially open a dialogue about how to include other concerns – human, food and energy security – into the climate change debate (Fein et al. 2010: 10–11).

Conclusions

Energy security and climate change policy are two of the most pressing policy issues facing governments today. While the two policy streams share many objectives, they are as likely as not to be in conflict. Climate change campaigners who advocate the end of coal neglect the military and economic security risks associated with increased reliance on natural gas and oil, particularly in the short to medium-term period before renewable technologies are available to scale. On the other hand, policy makers who neglect climate policy in favour of a one-eyed quest for energy security run the risk of provoking an environmental disaster that carries economic, military and human security risks of tremendous magnitude. Wherever possible, policy makers must look for solutions that bridge the divide between energy security and climate policy. There is near universal consensus that energy efficiency provides this vital link, but it has so far been severely under-deployed in both the climate policy and energy security realms.

Energy efficiency measures have benefits for both the developed and developing world, but the pay off is greatest in the rapidly industrialising countries where energy demand is growing fastest and energy intensity is relatively high. The Clean Development Mechanism (CDM) of the Kyoto Protocol is one of the only multilateral mechanisms available to facilitate the transfer of low-carbon technology from the developed to the developing world, but it has so far failed to deliver substantial results in the area of energy efficiency. The future shape of the mechanism is in doubt with the continued negotiations on a post-2012 successor agreement to the Kyoto Protocol. Whatever form the revised CDM takes, it must play a greater part in scaling up energy efficient technologies to the rapidly developing economies that are currently poised to be the next great energy consumers of the twenty-first century. The CDM has the opportunity to bridge the gap between energy security and climate policy in bringing about the energy efficiency revolution the world so desperately needs.

Notes

1 Britain's former Prime Minister, Tony Blair, used this phrase in a speech to the United States on October 20, 2006. For more background see Hartley and Medlock (2008).

2 See Hamilton (1983) for the original discussion of the relationship between high oil prices and recession in the United States since World War II, and Yergin (1991) for a complete review of the strategic significance of oil.

3 There has been some discussion among United Nations Climate Change Conference party members of other targets, including 550 ppm CO2e and 350 ppm CO2e, this chapter will stick with the 450 ppm CO2e target identified by the International Panel on Climate Change and the International Energy Agency.

4 See The Stern Review 2006; the United States Department of Defense also highlighted the strategic vulnerabilities created by the US military's heavy reliance on imported oil, as well as the risks associated with climate change in its Quadrennial Defense Review Report. See United States Department of Defense (2010).

5 The International Energy Agency (IEA) estimates that US$10.5 trillion (2008-dollars) in new energy investment will be needed between 2010 and 2030 to keep the world on a path to 450 ppm CO2e (IEA 2009: 167), mostly in transport, building, electricity and industry.

6 In particular, the United States and China, both of which have shown reluctance to entering the global climate change framework. Studies have also pointed out that developing countries with limited resource endowments (e.g. Japan and much of the European Union) have been the leaders the global climate change negotiations while countries with greater resource endowments (the United States, Canada, Australia) have been the laggards. See Hartley and Medlock (2008L 1–5) and von Hippel et al. (2009: 3–4).

7 Russia has shown its willingness to use its natural gas reserves as a form of leverage over neighbouring countries. For a discussion of the 2009 Russia-Ukraine gas dispute see, IEA (2009: 121).

8 For example, the promotion of energy efficiency is complicated by public ownership of power stations in many countries, where state revenue is dependent to some degree on energy consumption. Also, consumers have shown demonstrate their willingness to purchase vehicles with low fuel efficiency, even in the face of rising energy costs, so

behavioural factors cannot be discounted. For energy-exporting countries, the dynamic is further complicated by those states' overwhelming dependence on fossil fuel revenues. It is beyond the scope of this chapter to discuss these barriers in depth, but they cannot be overlooked in any global push to scale-up energy efficiency measures.

9 China – the world's largest emitter of GHGs – has announced its intention to reduce energy intensity by 20 percent by 2015. But in the first quarter of 2010, sales of oil and coal-fired electricity soared 24 percent on year, prompting a special cabinet meeting to discuss the problem. See Bradsher (2010) for more details.

10 See Gupta and Ivanova (2009) for a discussion of the need for a global energy efficiency framework, also citing a lack of research in this general area.

11 Carbon dioxide, methane, nitrous oxide (NO2), hydrofluorocarbons (HFCs), perfluorocarbons (PFCs), and sulfur hexafluoride.

12 Specific targets vary from country to country.

13 Kyoto Protocol to the United Nations Framework Convention on Climate Change, 10 December 1997, 37 I.L.M. 22 (1998), Art. 12.

14 Kyoto Protocol to the United Nations Framework Convention on Climate Change, 10 December 1997, 37 I.L.M. 22 (1998), Art. 3.

15 Statistics available at: http://cdm.unfccc.int [accessed 18 May 2010].

16 Project applicants can wait an average of more than 300 days for validation from a third-party auditor and another 200 days for Executive Board certification (Capoor and Ambrosi 2009: 39–40). To date, only half of the 4,000-plus projects waiting in the CDM pipeline have received final approval.

17 Kyoto Protocol to the United Nations Framework Convention on Climate Change, 10 December 1997, 37 I.L.M. 22 (1998), Art. 12.3(c).

18 See Ayers and Ayers (2009) for a complete discussion of the potential energy security and climate change benefit to capturing waste heat from fossil fuel powered manufacturing.

19 UNFCCC EB 22, Annex 3, "Additional Clarifications Regarding the Treatment of National/Sectoral Policies and Circumstances," November 2005.

References

Arquit Neiderberger, A. 2008. Scaling up energy efficiency under the CDM. In K. Holm Olsen and J. Fenhann (eds) *A Reformed CDM, Including New Mechanisms for Sustainable Development*. Copenhagen: UNEP Risø Centre, 127–45.

Ayres, R.U. and Ayers, E.H. 2009. *Crossing the Energy Divide: Moving from Fossil Fuel Dependence to a Clean Energy Future.* New Jersey: Wharton School Publishing.

Barker, T., Bashmakov, I., Bernstein, L., Bogner, J., Bosch, P., Dave R., et al. 2007. Technical summary. In B. Metz, O. Davidson, P. Bosch, R. Dave and L. Meyer (eds) *Climate Change 2007: Mitigation. Contribution of Working Group III to the Fourth Assessment Report of the Intergovernmental Panel on Climate Change*. Cambridge: Cambridge University Press, 26–93.

Blyth, W. and Lefèvre, N. 2004. *Energy Security and Climate Change Policy Interactions: An Assessment Framework*. IEA Information Chapter. Paris: OECD/IEA.

Boyd, E., Hultman, N., Timmons Roberts, J., Corbera, E., Cole, J., Bozmoski, A., Ebeling, J., Tippman, R., Mann, P., Brown, K. and Liverman, D. 2009. Reforming the CDM for sustainable development: Lessons learned and policy futures. *Environmental Science and Policy*, 12, 820–31.

Bradsher, K. China's energy use threatens goals on warming. *The New York Times*. [Online article dated 6 May 2010]. Available at: http://www.nytimes.com/2010/05/07/business/energy-environment/07energy.html [accessed 1 June 2010].

Capoor, K. and Ambrosi, P. 2009. *State and Trends of the Carbon Market 2009*. Washington, DC: World Bank. Available at: http://wbcarbonfinance.org [accessed 18 May 2010].

Carbon Trust. 2008. *Global Carbon Mechanisms: Emerging Lessons and Implications*. Available at: http:///www.carbontrust.co.uk.

Chester, L. 2010. Conceptualising energy security and making explicit its polysemic nature. *Energy Policy*, 38, 887–95.

Energy Information Agency. 2006. *International Energy Annual 2006*. 19 December 2008. Available at: http://www.eia.gov [accessed 1 April 2010].

Expert Group on Energy Efficiency. 2007. *Realizing the Potential of Energy Efficiency: Targets, Policies, and Measures for G8 Countries*. Washington, DC: United Nations Foundation.

Fein, I., Matsumoto, H., McNish, T., and Miller, J. 2010. Clean development fund: A 'public option' for carbon offsets. *Harvard Law and Policy Review Online*, 4. 10 February 2010. Available at: http://hlpronline.com/2010/02/clean-development-fund-a-%E2%80%9Cpublic-option%E2%80%9D-for-carbon-offsets/ [accessed 17 May 2010].

Figueres, C. 2006. Sectoral CDM: Opening the CDM to the yet unrealised goal of sustainable development. *McGill International Journal of Sustainable Development Law and Policy*, 2, 5–25.

Gupta, J. and Ivanova, A. 2009. Global energy efficiency governance in the context of climate politics. *Energy Efficiency*, 2, 339–52.

Hamilton, J.D. 1983. Oil and the macroeconomy since World War II. *Journal of Political Economy*, 25(2), 75–96.

Hartley, P. and Medlock, K. 2008. *Climate Policy and Energy Security: Two Sides of the Same Coin?* Working Chapter Series – The Global Energy Market: Comprehensive Strategies to Meet Geopolitical and Financial Risks. James A. Baker III Institute for Public Policy of Rice University. Available at: http://www.rice.edu/energy/publications/WorkingChapters/IEEJClimatePolicy.pdf [accessed 7 June 2010].

Hinostroza, M., Cheng, C., Zhu, X. and Fenhann, J. 2007. *Potentials and Barriers for End-Use Energy Efficiency Under Programmatic CDM*. United Nations Environment Programme. Riso Centre. Energy, Climate and Sustainable

Development. Available at: http://figueresonline.com/publications/potential.pdf [accessed 18 May 2010].

IETA. 2008. *State of the CDM 2008: Facilitating a Smooth Transition into a Mature Environmental Financing Mechanism.* International Emissions Trading Association.

IEA. 2008. *Energy Efficiency Policy Recommendations: Worldwide Implementation Now.* Paris: OECD/International Energy Agency. Available at: http://www.iea.org [accessed 20 May 2010].

IEA. 2009. *World Energy Outlook 2009.* Paris: OECD/IEA.

IPCC. 2007. *Working Group III Report: Mitigation of Climate Change.* Intergovernmental Panel on Climate Change. Available at: http://www.ipcc.ch/ipccreports/ar4-wg3.htm [accessed 15 May 2010].

Klare, M. 2009. *Rising Powers, Shrinking Planet.* New York: Henry Holt and Co.

Kruyt, B., van Vuuren, D.P., de Vries, H.J.M. and Groenenberg, H. 2009. Indicators for energy security. *Energy Policy*, 37, 2166–81.

Lloyd, B. and Subbaro, S. 2009. Development challenges under the Clean Development Mechanism: Can renewable energy initiatives be put in place before peak oil? *Energy* Policy, 37, 237–45.

Michaelowa, A., Hayashi, D. and Marr, M. 2009. Challenges for energy efficiency improvement under the CDM: The case of energy-efficient lighting. *Energy Efficiency*, 2, 353–67.

Moran, D. and Russell, J.A. 2008. *Energy Security and Global Politics: The Militarization of Resource Management.* New York: Routledge.

Samaniego, J. and Figueres, C. 2002. Evolving to a sector-based clean development mechanism. In Kevin Baumert (ed.) *Building on the Kyoto Protocol: Options for Protecting the Climate.* Washington, DC: World Resources Institute, 89–108.

Stern, N. 2006. Stern Review: The economics of climate change. *Executive Summary.* Available at: http://webarchive.nationalarchives.gov.uk/+/http://www.hmtreasury.gov.uk/media/4/3/Executive_Summary.pdf [accessed 18 May 2010].

Sutter, C. and Parreno, J.C. 2007. Does the current CDM deliver its sustainable development claim? An analysis of officially registered CDM projects. *Climatic Change*, 84, 75–90.

UNEP. 2010. Risø Centre. Available at: http://www.cdmpipeline.org [accessed 7 June 2010].

United States Department of Defense. 2010. *Quadrennial Defense Review Report.* Available at: http://www.defense.gov/qdr/QDR%20as%20of%2029JAN10%201600.pdf [accessed 30 May 2010].

von Hippel, D., Suzuki, T., Williams, J., Savage, T. and Hayes, P. 2009. Energy security and sustainability in Northeast Asia. *Energy Policy* [doi:10.1016/j.enpol.2009.07.001].

Wara, M. 2008. Measuring the clean development mechanism's performance and potential. *UCLA Law Review*, 55, 1759–803.

Wenger, A., Orttung, R. and Perovic, J. 2009. *Energy and the Transformation of International Relations: Toward a New Producer-Consumer Framework.* Oxford: Oxford University Press.

Yergin, D. 1991. *The Prize: The Epic Quest for Oil, Money and Power.* New York: Free Press.

Chapter 5

An Economic and National Security Perspective on Critical Resources in the Energy Sector

Nayantara D. Hensel

Over the past 50 years, global economic growth, combined with greater specialization of national economies, has strengthened the web of global economic interdependence and contributed to an increased flow of capital and goods between countries. As the global economy has expanded and the world's population has grown, the strain on key natural resources has dramatically increased, with the demand for natural resources often exceeding supply. As a result, many national economies have become increasingly dependent on critical natural resources located abroad. This, in turn, can lead to national security concerns, especially if political instability in the natural resource-rich countries threatens dependable export flows of these critical natural resources. In response to these national security concerns, countries that are reliant upon imports of critical raw materials often pursue several strategies. These include:

a. diversifying the countries from which key resources are imported;
b. undertaking capital investments to extract key resources domestically, thereby reducing dependence on imports; and
c. developing substitutes or alternatives for key resources.

The purpose of this chapter is to analyze and assess these strategies in the context of the energy sector within the broader US economy by examining the degree to which the US is dependent on other countries for its energy needs and evaluating the degree to which:

a. the US has pursued development of alternative energy resources to reduce overseas dependence on critical raw materials;
b. the feasibility of capital investments in developing domestic sources for these critical resources; and
c. the risks and rewards of developing alternative energy options.

The Biofuels Sector

The dependence of the US economy on petroleum imports from Middle Eastern countries has traditionally been an area of national security concern, both due to the resulting dependence of key US activities on inputs from overseas countries, as well as due to the possibility of political instability in some of the countries. In response to these concerns, the US has diversified its sources of crude oil imports. Data from the Energy Information Agency indicates that US imports of crude oil and petroleum products from OPEC countries have fallen from 50.5 percent of all crude oil and petroleum product imports in 1993 to 41.7 percent in the first ten months of 2010. Indeed, imports from Saudi Arabia, which constituted 16.4 percent of US crude oil and petroleum imports in 1993, had fallen to 9.17 percent in the first ten months of 2010. Although imports from Iraq have increased as a percentage of total crude oil imports into the US between 1993 (0.8 percent) and 2010 (3.6 percent), they have fallen from their peak during this period of 6.7 percent in 1999. Kuwait's share has fallen, as well, over the period, from 4 percent in 1993 to 1.7 percent in 2010. The US has also reduced its imports from some other oil producing countries: crude oil imports from Venezuela fell from 15.1 percent in 1993 to 8.4 percent in 2010 and imports from the UK declined from 4.1 percent to 2.23 percent of total US crude oil imports during the same period. Since 1993, imports from Russia have increased significantly, from zero to 5.2 percent, as have imports from Canada and Brazil.

Diversification of countries from which the US imports crude oil is just one of several strategies. Over the last ten years, there has been an increasing emphasis on biofuels as a means of reducing pollution, as well as a means of minimizing dependence on foreign energy sources. The US has experienced rapid consumption and production of biofuels since 2000 and is one of three countries which had an excess of production of biofuels over consumption of biofuels in 2009 (for the first time since 2002). Data from the Energy Information Agency indicates that US biofuel consumption was 46.8 percent of global biofuel consumption. Central and South America's biofuel consumption was about 27.7 percent of global biofuel consumption, while Europe's consumption was about 18 percent of global consumption. A similar pattern could be found on the production end: the US production of biofuels was 45.6 percent of global biofuel production, followed by Central and South America's biofuel production, which was about one third of global biofuel production, and then followed by Europe, whose biofuel production was about 14.3 percent of global production. US biofuel production and consumption have grown by about 550 percent between 2001 and 2009, which is significantly faster than the growth of global biofuel production and consumption globally (excluding the US). Brazil is second to the US in biofuel production and consumption, which are 55–65 percent that of the US, followed by France and Germany, which together produce and consume less than 20 percent of the US's biofuel production and consumption. Consequently, the US has expanded its universe of alternatives to petroleum to include biofuels—an area in which it is not significantly dependent on other countries since it produces as much as it consumes. Indeed, Brazil is the only country which has consistently run a substantive surplus in biofuel production over consumption.

Does the US, however, have enough biofuel production infrastructure to make biofuels a viable alternative source? Developing a substantive biofuels industrial base requires capital investments, as well as time. Renewable energy standards mandate that the American fuel supply must include 36 billion gallons of biofuels by 2022.[1] To achieve this, the US will need to build about 527 biorefineries, with an estimated cost of $168 billion.[2] The predicted capital cost for an alternative fuels refinery varies from $2–$12 per gallon of output.[3] Moreover, a second concern is the length of time that it takes to build the biorefineries and for them to reach maximum levels of production. After producers license the technology, a biofuels unit takes 24–30 months to become a viable operating unit. These constraints suggest that, despite its rapid growth in biofuels production over the past decade relative to the rest of the world, the US will only have a production capacity of a few hundred million gallons by the end of 2010, and about a billion gallons by 2015–2016. This represents a significant shortfall relative to the 36 billion gallon goal by 2022.[4]

A third concern regarding the biofuels industry is the high cost of biofuels relative to traditional, petroleum-based fuels. This can impose financial hardship on government agencies and companies, a consideration that is especially significant during periods of slow economic growth and fiscal austerity. The difficulties in adopting biofuels, as well as the benefits, are evident in the challenges and successes of the Department of Defense in the biofuels area.

The Department of Defense is the largest single consumer of oil in the world.[5] Indeed, in 2008, the US military's annual fuel consumption was approximately 200 million barrels.[6] About 75 percent of DoD's energy consumption is in "mobility" type fuel, which is used in ships and aircraft; jet fuel comprises over half of the "mobility" fuel consumed.[7] Consequently, both from a cost perspective and a national security perspective, the Department of Defense has a strong incentive to switch to a cheaper, domestically-produced fuel. The challenge is in the cost and time that it takes to develop the biofuels, which are included in the current prices paid, as well as the importance of finding biofuels that work in different types of ships and aircraft. The Navy, for example, uses at least seven different types of petroleum-based fuels and burns nearly 35 million barrels per year.[8] DoD has partnered with industry and with universities to develop biofuel processes and technologies. For example, UOP, a subsidiary of Honeywell, developed a bio-based fuel composed 50 percent of a petroleum-based fuel and 50 percent of a jatropha and algae-based biofuel—a research effort which was funded by the US Defense Advanced Research Projects Agency (DARPA).[9] Using algae-based fuels and camelina jet biofuels can reduce carbon emissions by 80 percent compared with conventional kerosene.[10] Similarly, Purdue has worked with NASA, Penn State, and the Air Force Office of Scientific Research to develop an aluminum ice fuel to launch a rocket.[11]

The costs of biofuels, however, are high because they incorporate research and development/start-up costs and are not yet produced in large volumes. This can lead to higher costs than traditional fuels, which is difficult in the current period of fiscal austerity and slow economic growth. DoD will pay $2.7 million for 40,000 gallons of jet biofuel from camelina at $67.50 per gallon.[12] Since this includes R&D, as well as set-up costs, this could fall over time. Similarly, algae-derived fuel for use on ships, produced by Solayzme, will cost about $424 per gallon.[13] Finally, in order for biofuels to be a viable alternative to traditional fuels, thus reducing the dependence of the US on other countries, contracts with the Department of Defense need to be of longer duration, thereby encouraging lenders and venture capitalists to invest in this sector. Currently, the Defense Energy Support Center (DESC) can only award contracts for up to five years to purchase fuel. Attractive investment returns for lenders, however, are more likely to be achieved with 15–20 year contracts. [14]

The Navy's efforts to test the usage of biofuels shows substantial progress, but significantly more usage and planning must occur for biofuels to be a viable alternative to traditional fuels. The Navy plans to have a green strike group by 2012, and a Great Green Fleet by 2016. By 2015, the Navy plans to consume half of the petroleum in its non-tactical vehicles that it currently uses and, by 2020, 50 percent of energy is to come from alternative sources. By 2020, the Navy will need 8 million barrels of biofuels—4 million barrels of F-76 military diesel fuel and 4 million barrels of JP-5 jet fuel. [15] On April 22, 2010, the Navy's first flight to break the sound barrier using biofuels took place on one of the Navy's fighters, the F/A "Green" Hornet, which used the 50/50 mix of camelina-based biofuel made by UOP.[16] In June, 2010, the Royal Netherlands Air Force used this same blend in a test flight of the AH-64D Apache, which was the first helicopter flight using a biofuel, while the Air Force used a biofuel blend on its test flight in March, 2010 with the A-10 Thunderbolt.[17] The first commercial flight using biofuels occurred in April, 2010, when United Airlines used them on an Airbus 319. [18]

Consequently, while the US has explored alternative energy sources to traditional petroleum-based fuels, such as biofuels, and has expanded both its consumption and production rapidly over the past ten years relative to the rest of the world, significant challenges remain to be addressed before the promise of biofuels can be fully realized. Although biofuels reduce the critical dependence of the US on other countries for energy sources, the US will need time and significant capital investment to build the biofuel industrial base and to develop cost-effective fuels.

Rare Earth Minerals

Uses of Rare Earth Minerals

In addition to the development of biofuels, the US has placed a greater emphasis on the development of other alternative technologies in an effort to reduce US

Table 5.1 Distribution of Uses for Rare Earth Minerals

Use	Percentage of Total Volume Used
Metal alloy	29%
Electronics	18%
Chemical catalysts	14%
Phosphors	12%
Catalytic Converters	9%
Polishing and Ceramics	6%
Magnets	5%
Oil refining catalysts	4%

Source: US Department of Energy's *Critical Materials Strategy*, December, 2010, p. 38.

Table 5.2 Uses of Specific Rare Earth Elements

Element	Magnets for Vehicles	Magnets for Wind Turbines	Batteries for Vehicles	Lighting
Yttrium				X
Dysprosium	X	X		
Terbium				X
Europium				X
Samarium	X	X		
Neodymium	X	X	X	
Praesodymium	X	X	X	
Cerium			X	X
Lanthanum			X	X

Source: US Department of Energy's *Critical Materials Strategy*, December 2010, p. 14.

dependence on oil from the Middle East and other overseas countries. Nevertheless, the US may be substituting dependence on one group of countries for dependence on another area of the world since many of these alternative technologies require rare earth minerals as an input and China controls 95–97 percent of the production of these critical minerals.

Rare earths elements (of which there are 17) can be found in iPhones, Blackberries, wind turbines, hybrid cars, lasers, magnets, fiber optics, computer disk drives and memory chips, LCD's, superconductors, X-ray tubes, catalytic converters, rechargeable batteries, fluorescent bulbs, the electric motors on the guidance fins of missiles, etc.[19] Table 5.1 shows the uses of rare earth minerals by volume and highlights their usage in metal alloys, electronics, catalysts, phosphors for lighting, catalytic convertors, and magnets.

Table 5.2 shows the uses of specific rare earth elements; many of them are valuable inputs into clean energy technologies, such as magnets for wind turbines and vehicles, batteries, and energy efficient lighting. A shortage of one of these elements can lead to production disruptions in a variety of different products.

The use of rare earth minerals in the electric motors in fuel efficient cars is particularly important since, in addition to environmental considerations, the move toward fuel efficient vehicles is partially motivated by a desire to reduce dependence on foreign oil. About 180,000 Toyota Priuses will be sold this year in the US and each one uses 2.2 pounds of neodymium, which is mainly found in abundance in China.[20] Neodymium is a vital component to electric vehicle motor magnets and terbium and dysprosium are added to neodymium to allow it to remain magnetic at high temperatures.[21] About 30 companies have hybrid electric, plug-in hybrid, and battery electric vehicle programs underway.[22] Indeed, due to usage by electric vehicles, dysprosium, neodymium, praseodymium, etc. demand is set to grow by an average of 790 percent during the next five years.[23] Large capacity wind turbines also use motors with magnets composed of neodymium, iron, and boron. Wind turbines use several hundred kilograms of neodymium, while electric drive vehicles use up to a kilogram.[24]

Based on the potential new CAFÉ requirements, it is possible that over half the cars sold in the US by 2025 will be hybrid cars, which will create a significant demand for rare earth minerals. Indeed, the market for electric and hybrid cars may reach $100.5 billion by 2014–2015.[25] Toyota plans to have 11 redesigned or new hybrids by 2013.[26]

One of the issues concerning the future expected demand for rare earth minerals is whether the market for hybrid cars and other alternative energies will actually attain the anticipated size. Will the up-front costs of a hybrid outweigh its longer-term efficiency benefits, thus reducing actual future demand for hybrids relative to forecasted demand? If the up-front costs outweigh the benefits, then the demand for rare earth minerals needed for the batteries of hybrid vehicles may be less than anticipated. For example, hybrid cars often have higher costs of repairing engines or collision damage repairs than a non-hybrid car.[27] Hybrid cars are more expensive than conventional models by about $3,000–$5,000, while the Volt, the Leaf, and other plug-in cars cost between $10,000 and $20,000 more than conventional cars.[28] A 2010 Honda Civic hybrid model costs $3,545 more than a non-hybrid model; it is unclear if the few hundred dollars in annual gas savings offsets these additional costs.[29]

Although the goal of the Obama administration is to have 1 million electric cars in use by 2015, and $11 billion has been allocated to develop the technology, the low market share for hybrids suggests that this goal may not be reached. According to an October 27, 2010 report by J.D. Powers and Associates, sales of electric and hybrid cars in the US will be 2.2 percent of global sales of electric and hybrid cars. J.D. Powers projects that electric and plug-in hybrids will comprise 7.3 percent of the car market by 2020.[30] Sales of hybrids in the US have actually fallen for the past three years. Sales of the Toyota Prius made up half of the sales this past year and US government purchases made up about a quarter of the sales of hybrids made by Ford and General Motors. About 10 percent of total purchases of hybrids over the past two years were made by the General Services Administration, which purchased 14,584 hybrid vehicles for the government fleet, of which funds for 3,100 of them came from the 2009 stimulus package.[31]

Broader studies beyond the hybrid market have suggested that the benefits of improved cost efficiencies from energy efficient technologies may be overstated. Installation of solar electricity in a house costs $18,000; do reduced energy bills and tax savings offset this? For high efficiency washing machines, the difference is similarly tight. Whirlpool's high efficiency machines save water bills by $15 per year, but cost about $70 more than the non-high efficiency counterpart,[32] which suggests that it would take about five years to break even.

Government funding is a significant factor in the development of renewable energy resources. On the one hand, renewable energy resources reduce the dependence of the US on oil; on the other hand, some of the renewable technologies can increase the dependence of the US on rare earth minerals. Renewable energy sectors, such as solar energy, wind, and biofuels are highly dependent on government funds. Although the stream of investing has been volatile in the alternative energy sector, corporations have invested about $5.1 billion in clean energy programs in 2010, which is a 325 percent increase over the previous years. The role of the government has become more important as venture capital and private equity firms have reduced their involvement in terms of the number of companies that they are backing; venture capitalists did, however, invest $4.02 billion in companies involved with clean energy during the first half of 2010, which is a 43 percent increase from the first half of 2009. As the government continues to back new technologies, technologies which would otherwise become too inefficient or expensive to be widely adopted are supported; indeed, although solar energy only accounts for less than 1 percent of all the energy in the world, it accounts for 50 percent of renewable energy financing from the government.[33] One example of government support for renewable energy has been Section 1603 of the American Recovery and Reinvestment Act, which includes the US Treasury's renewable energy tax grant program. This program compensates for up to 30 percent of the cost of alternative energy projects after they are built. The grant program was an important factor in developing 1,170 solar projects in 42 states through dispensing $1.3 billion, as well as in assisting 211 wind power efforts across 38 states through expending $15 billion.[34] Elimination of programs like these due to fiscal austerity would result in larger companies being likely to continue to invest in wind energy and other alternative energies, but smaller companies being less able to continue to afford doing so.[35]

Rare earth minerals also have defense applications: the motors that run the tail fins and rudder on the F-22 use magnets that involve lightweight rare earths. Neodymium magnets are used for hybrid electric motors under development to conserve fuel for Arleigh Burke-class destroyers.[36] Yttrium is used for laser gun sights in the General Dynamics Abrams tank. Lockheed Martin's radar on Aegis destroyers contains samarium-cobalt magnets that will need to be replaced over 35 years.[37] Motors in weapons like the JDAM would be three times larger without advanced magnets containing neodymium, which replace a heavier, more expensive hydraulics system.[38] Neodymium is a key input to solid-state lasers which select targets, and samarium-cobalt permanent magnet motors are in the fins that steer precision bombs.[39]

Rare earth minerals are also used in the new, energy-efficient lights. The newer generation of lighting uses phosphors that contain rare earth elements, such as yttrium, terbium, europium, lanthanum, and cerium.[40] White LED (light-emitting diodes) use cerium and europium phosphors, rather than lanthanum and terbium phosphors. Neodymium is used in very small percentages. It is likely that the use of rare earth elements in phosphors will increase for linear fluorescent lamps and compact fluorescent lamps.[41] Indeed, the June, 2009 guidance issued by the Department of Energy has phased out most non-rare earth linear fluorescent lamps by requiring increased efficiency, as measured by lumens per watt.[42] The compact fluorescent lights are manufactured almost entirely in China, although linear fluorescent lamps are assembled in Europe and North America.[43]

Catalytic converters use cerium to reduce carbon monoxide emissions. Due to the importance of catalytic converters in passenger vehicles, cerium's use in catalytic converters composed 9 percent of 2008 rare earth element consumption.[44]

Finally, another use of rare earth elements is in the fluid cracking catalysts inherent in the conversion process of producing gasoline, distillates, and lighter oil products from heavy oils. Lanthanum is used in the process, along with cerium and neodymium to a lesser degree. Although rare earth elements used for the catalysts make up only a small percentage of the cost of petroleum products and are not requirements for refining, an absence of rare earth elements would affect the yields of refineries and result in greater capital investments to re-develop conversion processes which do not need rare earth elements.[45]

Despite the dependence of the US on China for rare earths, such as neodymium, it is important to note that the patents and intellectual property rights for the technologies using the rare earth minerals are held by other countries besides China. Consequently, although in developing technologies relying on rare earths, the US is increasing its dependence on China for the inputs, control of the intellectual property rights is much more diversified. For example, although the patents on neodymium-iron-boron magnets are held by Magnequench (which, although formerly a US-owned firm, has been owned by a Chinese-based consortium since 1995) and Hitachi Metals (Japan), the licenses for producing sintered neodymium-iron-boron magnets are held by 10 firms in Germany, Japan, and China.[46] Similarly, despite the significant involvement of Chinese firms in the manufacturing process of lighting, US firms still hold the intellectual property rights for the formulas behind the fluorescent lighting phosphors which enables them to have more control over the R&D phase in the value chain.[47] Consequently, one strategy for mitigating dependence on other countries for critical materials is to control the means of using those inputs in the value chain by retaining intellectual property rights, despite the dominance of the other countries in having the reserves of the input and in dominating key input production.

Sources of Rare Earth Minerals

Although China mined 120,000 tons of rare earth minerals in 2008, followed by India (2,700 tons), Brazil, and Malaysia, China has only 37 percent of rare earth reserves.[48] The US has the second largest concentration at 13 percent, Russia and Australia have 5–6 percent of the known reserves, while India, Canada, and Greenland have appreciable quantities.[49] Weak environmental regulation and cheap labor, however, allowed China to undercut global producers in the 1990s and made mining sufficiently unprofitable that many non-Chinese firms went out of business.[50] Thorium, a radioactive element, is often found in the same ore as rare earth minerals (with the exception of some mines in southern China), which makes mining environmentally hazardous.[51]

Low Chinese prices and the cleanup from a radioactive leak contributed to the closure of Molycorp, the only American rare earth mineral mining facility, in 2002.[52] Molycorp's mine, Mountain Pass, is in the Mojave desert and was the primary source of rare earth minerals in the US, in addition to comprising about 50 percent of global production at one time.[53] Production of rare earths in Mountain Pass began in the mid 1960s and the facility was the leader in rare earth mineral production globally until 1984. Rare earth mining stopped in 2002 and, as of December 2010, has been focused on reprocessing the stockpile of rare earth ores at Mountain Pass.[54]

Exploration of other areas rich in rare earth minerals is one solution, although, due to high capital investment, external funding is important. The biggest mines under development outside China are Great Western Minerals in South Africa; Avalon Rare Metals in Nechalacho, Canada; Molycorp in Mountain Pass, CA; Lynas Corporation at Mount Weld in Australia; and Arafura Resources in Nolans, Australia.[55] Avalon Rare Earths requires $792.6 million to bring the Nechalacho mine into production, which has concentrations of terbium and dysprosium.[56] Molycorp plans to reopen its mine in the US with better technologies and hopes to begin producing 20,000 metric tons of rare earth oxides annually by 2012, which is one sixth of global production and more than the entire demand in the US. Consequently, Molycorp plans to supply a portion of the demand from Japan and the European Union.[57] Nevertheless, if the US does successfully move toward increased usage of electric and hybrid cars or increased usage of solar and wind power, then demand for rare earth minerals in the US may exceed supply.[58] Molycorp raised $379 million in its IPO in August, which will be used to build more facilities. Since the project will cost $511 million, Molycorp will need to raise an additional $150 million through vendor finance from sellers of equipment, borrowing from commercial banks, or US government funds.[59] Molycorp and the Australian firm, Lynas, hope to jointly produce 40,000 tons of rare earth minerals annually by 2012, which is about 35 percent of China's current tonnage.[60] By 2013, Great Western, which is located in Saskatchewan, plans to develop the Steenkampslraal mine in South Africa in order to compete with the Chinese.[61] Concerns have arisen, however, that, as rare earth mineral production outside of

China increases, price wars could ensue and mining companies outside of China could, again, find it unprofitable to mine and have to close operations. Indeed, when China flooded the tungsten market in the early 1990s, the lower prices led to manufacturers closing operations due to lack of profitability.[62]

Several Australian rare earth mines have been under development for the past eight years, yet, since Australian production will have greater environmental protections and higher labor costs than Chinese production, extraction of its rare earth minerals will likely be more expensive than those from China. In addition, it is possible that, in the time that it takes to develop rare earth mineral mines, alternatives to the use of rare earth minerals may develop.[63] Australia's Arafura rare earth mine plans to produce 22,000 tons annually of rare earth minerals beginning in 2013, which will only be small step toward alleviating China's dominance in rare earth mineral production.[64]

In addition to its domestic production of rare earth minerals, China continues to expand its control of rare earth mines outside of China, although China has not allowed purchase of minority or majority stakes by foreigners in the rare earth sector within China. Over the past five years, Chinese companies have made three attempts to acquire Molycorp in the US.[65] Indeed, Molycorp was owned by Unocal, for whom the China National Overseas Oil Corporation (CNOOC) bid in 2005. Due to concerns that China might control important assets in the oil industry, Chevron purchased Unocal. Molycorp was sold in 2008 to private equity firms, including Goldman Sachs and Resource Capital Funds.[66] Moreover, East China Exploration, a Chinese state-owned enterprise, acquired 25 percent of Arafura last year, while the attempt of a Chinese enterprise to acquire 52 percent of Lynas Corporation, which plans to open a rare earth mineral mine in Australia in 2011, was blocked by Australia on national security grounds.[67] Similarly, following the discovery of a rare earth deposit in Australia by Northern Uranium of Australia, a Chinese state-owned company attempted to purchase a 51 percent interest in it, which was declined. Subsequently, the Chinese company purchased a 13.5 percent interest in it on the open market.[68]

The newest source of rare minerals, as recently re-discovered, is in Afghanistan, which has nearly $1 trillion in untapped mineral deposits.[69] With the right infrastructure and political stability, it could become one of the most important mining centers in the world.[70] About three quarters of the value of the deposits is in iron and copper, with a much smaller share for rare earths.[71] Lithium deposits may be as large as those in Bolivia, which has the world's largest lithium reserves (50 percent of the world's supply).[72] Deposits are located throughout the country, including the southern and eastern regions and along the border with Pakistan in combat zones.[73] It will take years, lacking infrastructure or mining industry, for Afghanistan to fully exploit its mineral wealth.[74] There are concerns, however, that China may try to gain market share in the Afghan mineral deposits (China won the bid for the Aynak copper mine in Logar Province).[75]

The economic benefits of mining for rare earth minerals are dependent on the concentrations of light rare earth minerals (neodymium, lanthanum, cerium, praesodymium), heavy rare earth minerals (yttrium, dysprosium, terbium, holmium, erbium, thulium, ytterbium, lutetium), and medium rare earth minerals (europium, samarium, and gadolinium), since this has an impact on the cost of refining the rare earth minerals and the degree to which there is demand for these various types of elements. Thorium, a radioactive material, is in most rare earth deposits and, therefore, has to be separated out very carefully.[76]

Light rare earth minerals are found in the largest concentrations in Xunwu in the Jiangxi Province in China, in Bayan Obo in Inner Mongolia, and in Lovozerskaya in Russia. Medium rare earth minerals are found in the largest concentrations in Xunwu in the Jiangxi Province in China and in Longman, in the Jiangxi Province in China. Heavy rare earth minerals are also found in Longman in the Jiangxi Province in China and in Lahat, Perak, Malaysia.[77]

Over the next five years, several facilities are expected to become active producers, all of which have high concentrations of light rare earth minerals, but low concentrations of heavy rare earth minerals. These include the Molycorp mine in Mountain Pass, California, in the United States; Mount Weld, Australia; the east coast of Brazil; Nolans' Bore in Australia; Nechalacho in Canada; Dong Pao in Vietnam; Hoidas Lake in Canada; and Dubbo Zirconia in Australia. It is estimated that the Molycorp facility in Mountain Pass, CA largely contains cerium and lanthanum, some neodymium and praseodymium, and small concentrations of samarium, europium, gadolinium, and yttrium.[78]

Of the subset of new facilities being developed, the Nechalacho mine in Canada has the greatest concentrations of heavy rare earth minerals, especially yttrium. Several sites with rare earth minerals may be developed in the longer term; these include: the US Rare Earths Lemhi Pass quadrangle, in Idaho and Montana (US); Nangdang, in Guangdong in China; the east coast of Brazil; North Capel in western Australia; North Stradbroke Island in Queensland, Australia; and Green Cove Springs in Florida in the US. These facilities have higher concentrations of heavy rare earth minerals than the facilities which are coming online in the longer term. The Lemhi Pass quadrangle in Idaho and Montana has high concentrations of yttrium, cerium, and neodymium (in that order), followed by samarium and gadolinium. The Green Cove Springs area in Florida has high concentrations of cerium, followed by lanthanum and praseodymium.[79]

The Impact of Dependence on Rare Earth Minerals

The impact of the China's recent export restrictions and quotas on the price of rare earth minerals and the strategic behavior of the US and other countries highlights the risks of global dependence on China's supply of extremely important raw materials.

China's behavior of tightening export quotas of rare earth elements began in 1999, such that exports fell from 65,000 tons in 2005 to 50,000 tons in 2009. The most recent quota reduced exports by 40 percent, from 50,145 tons in 2009 to 30,258 tons for 2010.[80] China reduced its export quotas in July by 72 percent for the second half of the year, such that only 7,976 tons of rare earth minerals could be exported during that period.[81] China has suggested that increasing export quotas helps to reduce environmental damage in China and to preserve rare earth minerals for the expanding consumption needs of its own population.[82] Indeed, China is projected to require 54 percent of the rare earths which are mined next year, with Japan requiring 24 percent, followed by Europe at 10 percent, the US at 8 percent, and other countries at 4 percent.[83] China has also been concerned about uncontrolled mine development, which, they argue, has reduced China's untapped reserves from over 40 percent of global reserves in 1998 to one third or less of the global total.[84] In addition, as export restrictions increase, foreign companies are more likely to move the portion of their operations which is dependent on rare earth minerals to China to manufacture the necessary components. These reductions in export quotas have resulted in price increases.[85] Indeed, cerium oxide (the lowest value rare earth mineral) has increased 930 percent between 2007 and September, 2010 to over $35/kilogram. Dysprosium has increased 220 percent to $286/kilogram.[86] Critics of China's trade strategy of raising export taxes and quotas on rare earths, as well as other products, argue that the policies violate the World Trade Organization rules: when China joined the WTO in 2001, it promised to place export taxes on only 84 product categories, yet, as of 2010, China had taxes on 329 product categories.[87]

On September 21, 2010, as a result of a collision of Japanese and Chinese ships in disputed waters and Japan's detention of the Chinese fishing boat captain, China placed a de-facto embargo on exports of rare earth minerals to Japan, the largest rare earth importer.[88] This embargo ended in late November, 2010.[89] Embargoes were also placed for a limited time on shipments of rare earth minerals to the US and the EU.[90] China did not officially acknowledge that it had imposed a ban, such that the halt in exports could be construed as based on decisions of individual firms. This was important in that Japan would have a harder time protesting the export restriction before the World Trade Organization if the export restriction were not officially recognized by the Chinese government. The argument against the export ban being based on the decisions of individual firms is that of the 32 companies with licenses to export rare earths out of China, 10 of the firms are foreign companies, and it is unclear why foreign companies stopped shipments.[91] Like Japan, Germany also expressed significant concern about the interruption of rare earth exports from China since a significant number of German manufacturers are dependent on rare earths. The dependency of countries on these scarce resources could trigger diplomatic realignments, in that China and Germany have been in agreement concerning their lack of approval of the continuation of the Fed's quantitative easing program. Quantitative easing will weaken the dollar and make US exports cheaper, thus limiting demand for Chinese and German exports and reducing their respective trade surpluses. Germany and China have disagreed, however, concerning China's policy of tightening export restrictions of rare earth minerals due to the dependence of the German manufacturing sector on them.[92]

Concern over China's recent use of rare earth mineral export quotas and de-facto embargoes has encouraged countries to search for alternative sources, either by streamlining production processes to use fewer rare earth minerals, recycling, or identifying other sources of rare earth minerals besides China. [93]

At the end of November, the EU raised the possibility of more investment in transport, environmental, and energy infrastructure in Africa so that its rare earth mineral resources could be further explored.[94] The EU also planned to move toward free trade agreements with Latin American and Central American countries, as well as to engage in trade negotiations with Russia, Canada, and India.[95]

After China, Japan has the largest demand for rare earths, due to its production of cars and technologies requiring them as inputs. Japan is likely to need 32,000 tons next year, whereas all of the other countries combined (excluding China) will require 24,800 tons. If China keeps its export quota in 2011 at the level that it was in 2010 (around 30,000 tons), then Japan will be over 10,000 tons short of its needs.[96] Japan is negotiating agreements with Australia, Mongolia, and Vietnam to develop new mines.[97] Japan has also entered into an arrangement in which Sojitz Corporation will form a partnership with the Lynas Corporation (Australia) to distribute 80,000–90,000 tons of rare earth minerals mined by Lynas over ten years to Japanese markets.[98] Historians have argued that Japan's concerns over its lack of natural resources and its dependency on other countries have motivated its military actions: the US was a significant oil supplier to Japan in the 1930s and some argue that the decision of the US to impose an oil embargo on Japan in 1941 was a contributing factor in Japan's decision to bomb Pearl Harbor.[99]

The European Commission plans to invest 17 million euros for research into substitutes for rare earth minerals and also to improve technologies for mining more deeply underground; indeed, the European Commission estimates the value of unexploited rare earth minerals at depths of 500–1000 meters at 100 billion euros.[100] Nodules on the global seabed not only contain manganese, nickel, copper, and cobalt, but also rare earth minerals; this has been known for many years, but, due to the cost of mining on the seafloor, it has not been considered profitable, although it may become more lucrative with the increase in demand for rare earth minerals.[101] South Korea and Japan are forming a cooperative relationship on rare earth minerals and the development of rare earth processing technologies.[102] Vietnam and Japan are also partnering to develop rare earth mineral supplies.[103] Automakers of hybrids, such as the Toyota Prius, are shifting away from nickel metal hydride batteries, which require lanthanum, and towards lithium ion batteries.[104] Toyota plans to allocate $1.2 billion to engage in recycling and R&D for alternate materials, as well as to spend over half of the funds for overseas mining projects to improve access to rare earth minerals.[105]

Alternatives to the use of rare earth minerals in magnets are also being developed. For example, an electric motor using ferrite magnets has recently been developed by Hokkaido University and the New Energy and Industrial Technology Development Organization (NEDO) in Japan. Although these magnets will not be available for some years, they are significantly cheaper than magnets involving neodymium or other rare earths.[106]

In short, Japan's strategy in handling its dependence on rare earth minerals was intensified by the length of China's export restrictions, as well as the fact that, after China, Japan is the most significant consumer of rare earths. The Japanese government and Japanese companies have focused on recycling, making arrangements with alternative sources of rare earth minerals in other countries, and finding alternatives to rare earth minerals. On the other hand, due to the smaller relative manufacturing industrial base in the US for computers and other technologies, as well as for hybrid cars, American companies were less impacted by the export quotas and export restrictions than Japanese companies. The US defense manufacturing sector, however, is much more affected by the increase in prices of rare earth minerals relative to other sectors in the US economy, as are US consumers. Indeed, due to the tightening of the export restrictions, the price of a kilogram of samarium powder, which is used in the navigation system of the M1A2 Abrams Tank, made by General Dynamics, rose from $4.50 in June, 2010 to $34 in early September, 2010.

The US strategy in handling its dependence on rare earth minerals has been to consider ways to fund the re-opening of the only rare earth mineral mine in the US – Molycorp. On September 29, 2010, the US House of Representatives passed legislation to provide federal loan guarantees to domestic rare earth mineral producers.[107] As of February, 2011, the possibility of direct federal funding was being considered: Molycorp applied to the Department of Energy for $280 million.[108] Moreover, the Pentagon can provide financial support to companies to strengthen areas of the industrial base which are critical to national security under Title III of the Defense Production Act of 1950.[109] While the US has determined which other areas of the country have rare earth minerals, it has not been as aggressive as Japan in making arrangements with other countries for rare earth mineral supplies, largely due to a reduced need for rare earths in manufacturing relative to Japan. As discussed earlier, the role of governments in supporting rare earth mining in their countries may become more necessary if, after rare earth mines in other countries have begun production, China floods the market and the prices of rare earth minerals fall sufficiently far that, without government support, the nascent rare earth mining companies become unprofitable.

Other Non-Rare Earth Critical Elements

In addition to its dependence on foreign sources of rare earth minerals, the US is also dependent on other countries for a variety of non-rare earth mineral elements. These resources are equally important in energy efficient technologies, such as solar cells and batteries for vehicles. Table 5.3 summarizes the uses of some of these raw materials, such as lithium, cobalt, tellurium, gallium, and indium.

Table 5.3 Uses of Non Rare Earth Materials

Element	Solar Cells	Batteries for Vehicles
Lithium		X
Cobalt		X
Tellurium	X	
Gallium	X	
Indium	X	

Source: US Department of Energy's *Critical Materials Strategy*, December 2010, p. 14.

Many hybrid electric vehicles use nickel metal hydride batteries, which include rare earth elements—lanthanum, neodymium, praseodymium, and cerium—as well as non-rare earth elements—cobalt, nickel, manganese, and aluminum. PHEV's and electric vehicles have a greater need for batteries with higher storage capacity and higher power ratings than hybrid electric vehicles; as a result, they are more likely to use lithium ion batteries. Lithium ion batteries tend to have improved recharge rates, energy density, and performance in cold weather relative to nickel metal hydride batteries.[110] Growth in demand for lithium has been based on the development of the market for rechargeable lithium batteries, the production of which expanded by 25 percent between 2000 and 2007; indeed use of lithium in batteries has expanded from 6 percent of lithium usage in 2000 to more than 20 percent.[111] Cobalt consumption has been impacted by demand in the rechargeable battery sector and the super alloy sector, especially the manufacture of aircraft gas engine turbines—49 percent of the demand for cobalt in the US is based on aircraft gas turbine engine production.[112] Gallium consumption has been heavily influenced by the growth in demand for LED's used in computers and televisions; indeed, electronic components compose 98 percent of consumption of gallium. Integrated circuits made up 67 percent of gallium consumed, while optoelectronic technologies, such as LED's, photo detectors, solar cells, and laser diodes compose 31 percent of gallium consumption. The role of indium in films of indium-tin-oxide used for monitors, televisions, and phones has driven global demand for indium.[113]

Although China has 37 percent of the reserves of rare earth minerals and 73 percent of the reserves of indium, the US is dependent on other countries for other key materials. As Table 5.4 shows, Chile has 76 percent of the reserves of lithium and the Congo has 51 percent of the reserves of cobalt. Possible political instability in these countries could affect the supply and, therefore, the pricing of these materials. Similarly, about 50 percent of the world's cobalt reserves, as well as significant cobalt production, are located in the Democratic Republic of the Congo, which can lead to instability in supply sources due to the recent civil war in the region.

Production of some of these materials, on the other hand, is more diversified across countries. The largest suppliers of lithium are FMC (US), Tailson Minerals

Table 5.4 Distribution of Reserves of Non-Rare Earth Critical Materials

Countries holding reserves	Lithium	Cobalt	Indium
China			73%
Commonwealth of Independent States			
United States			3%
Chile	76%		
Argentina	8%		
Australia	6%	23%	
Peru			
Democratic Republic of Congo		51%	
Cuba		8%	
Others			3%

Sources: *Cobalt Facts*, Cobalt Supply and Demand for 2009. Cobalt Development Institute, 2009. "Critical Minerals and Emerging Technologies," *Issues in Science and Technology*, by P.G. Eggert, Summer 2010; *Mineral Commodity Summaries*, US Geological Survey (USGS), 1994–2010. US Department of Energy's *Critical Materials Strategy*, p. 30.

(Australia), SQM and Chemetall (both from Chile), and three mining companies in China. Chile produces about 41 percent of global lithium, while Australia produces 24 percent and China produces 13 percent.[114] Cobalt, which is produced in the process of mining for copper and nickel, is refined by China, Finland, and Canada, although the Congo hopes to refine cobalt again. Indium can be produced through mining zinc, as well as copper, lead, and tin ores.[115] China has the largest reserves of indium at 73 percent, with the US and Peru holding 3 percent and 4 percent of the reserves, respectively. Over half of the refining for indium occurs in China, with Japan and South Korea jointly refining 24 percent.[116]

For several of these key materials, the US discontinued production because, at the time, demand and pricing from foreign competitors made production unprofitable. For example, the US closed its mine in North Carolina in 1997 which produced lithium minerals and only has one facility involving lithium brine production in Nevada, since the South American mines were too difficult with which to compete.[117] Similarly, the US discontinued cobalt mining in 1971 and cobalt refining in 1985. Nevertheless, the US hopes to expand cobalt production in the US through projects in Idaho, Michigan, and Minnesota.[118] As demand for renewable energy increases and the technologies involving these inputs become more prevalent, the dependence of the US on other countries for these key materials will heighten and the pressure to re-open domestic mining and production will increase.

Table 5.5 shows the dependence of the US on imports of some of these key, non-rare earth materials. The US imports 91 percent of its rare earths from China. It is similarly dependent on China for 40 percent of its imports of tellurium and indium. The US, however, is dependent on Latin American countries for lithium

Table 5.5 Percentage Distribution by US Import Sources for Critical Materials

	Rare Earths	Lithium	Cobalt	Indium	Tellurium	Gallium
China	91%	1%	12%	40%	43%	16%
France	3%					
Japan	3%			19%		
Chile		63%				
Argentina		35%				
Norway			19%			
Russia			17%			
Canada				18%	18%	20%
Belgium					24%	
Germany						24%
Other	3%	1%	52%	23%	15%	40%

Source: US Department of Energy's *Critical Materials Strategy*, December, 2010, p. 38.

and imports 98 percent of lithium from Chile and Argentina. Despite the role of the Congo in producing cobalt, since refining takes place in other countries, the US has greater diversification in its cobalt imports. Nevertheless, for tellurium and for indium, the US is dependent on China for over 40 percent of its imports.

Conclusion

The web of interdependencies between countries is well illustrated in evaluating the national security implications of energy dependence of the US on other countries. Several strategies are often undertaken by countries to address national security concerns about the vulnerability to supply disruptions of critical resources. These include:

a. diversifying the countries from which the key resources are imported;
b. undertaking capital investments to extract the key resource within the home country to reduce dependence on imports; and
c. developing substitutes or alternatives for the key resource.

The US has attempted to increase its sources of domestic oil, as well as to diversify the sources of its imports of this strategic resource. The US has also exhibited the most substantive growth globally in consumption and production of biofuels, which do not increase its dependence on foreign sources of supply. Nevertheless, in order for biofuels to be viable alternatives to oil, biofuels production requires substantive capital investment in plants and facilities in order to generate the fuels in sufficient volumes. Moreover, although the US has funded substantive research and development for biofuels, the development of technology which can sustain the use of biofuels for an extended period of time as their primary means of energy, is key. In the current environment of fiscal austerity, the high costs of biofuels, which include research and development costs, make it difficult for cash-constrained companies and government agencies to develop long-term contracts with producers.

The dependence of the US, Japan, and other countries on China for rare earth minerals has become more significant over the past decade, as energy efficient technologies, such as wind turbines and hybrid cars, as well as consumer electronics, have developed which use these elements as inputs. Moreover, China has aggressively developed production in this sector, and, with its lower prices, made the few non-Chinese rare earth mineral producing firms unable to continue operations, thus reinforcing the dependence of other countries on Chinese resources.

The substitution of dependence on foreign oil with dependence on China for rare earth minerals will be a gradual process, which is highly dependent on the strength of the demand for energy efficient technologies, such as hybrid cars. China's tightening of its export quotas over the past several years has significantly increased the prices of rare earth minerals, which have especially impacted Japanese manufacturing. Japan's strategy at combating its dependence on rare earth minerals was much stronger and entrepreneurial than that of the US because the two countries have different compositions of their industrial bases. With the exception of defense manufacturing and some hybrid or wind turbine manufacturing, the US's exposure to rare earth mineral dependence on the production end is much weaker than that of Japan. As a result, the US has focused on reopening its earlier rare earth mines, while Japan, which, unlike the US, lacks the natural deposits, is forming arrangements with other countries which do have these deposits, identifying alternatives to rare earth minerals, and developing recycling programs. The EU, especially Germany, is following a similar strategy to that of Japan, due to the dependence of their manufacturing base on rare earths and their intrinsic lack of domestic deposits.

The US is also dependent on other countries, such as Chile and the Congo, for other critical inputs, such as lithium and cobalt. Political instability in these countries can disrupt their supply. Nevertheless, the mining operations often occur in a different country from the refining operations, which diversifies exposure. Moreover, in many cases, the intellectual property rights needed for the technology which uses these inputs are retained by the US or diversified from the country with the reserves of the input. This reinforces the interdependency, since, although the country with the reserves of the input can manipulate the price of the commodity for the US, the US and other countries control the means by which the demand for the input is generated, and have similar bargaining power in the bilateral relationship.

A significant parallel between the rare earth mineral sector and other sectors, such as lithium, is that the US had production operations in all of these sectors in the past, but closed these facilities since they could not compete with the other international producers. The resulting shift in the US industrial base away from manufacturing and towards services places a nation of consumers at the mercy of other nations with a stronger manufacturing orientation. If the US reopens these facilities as demand evolves, one concern is whether the re-opened facilities will be internationally competitive in the long-run. If not, government support may be necessary to strengthen sustainability of these sectors, since they are critical for national security. On the other hand, this may distort incentives for efficiency, innovation, and quality.

In conclusion, a country's endowment in natural resources does not always place it in a strong position internationally in that the industrial base of the country in mining and producing these critical inputs must be sufficiently well-developed that it can withstand international competition as demand evolves. The move toward energy efficient technologies and demand for non-petroleum resources has gained strength over the past few years, but it is unclear whether it will gain the anticipated sustained momentum in the coming years. The macroeconomic backdrop of fiscal austerity in the wake of budgetary and currency problems, as well as population growth and other demographic pressures, may have a significant influence on the magnitude and speed of the shifts in risk and exposure in the energy sector. Moreover, the web of interdependencies between nations has spread beyond critical materials and the energy sector, to include the financial sector, exchange rates, and monetary policy. The key to competitiveness under these shifting constraints may involve an understanding of the interrelationships between these forces and sustained flexibility in reshaping national strategies for sustainability.

Notes

1 Beidel, Eric. "Biofuels Industry: Friendlier Policies Needed to Displace Oil," *National Defense Magazine*, July 2, 2010.

2 Ibid.

3 Epstein, Curt. "Alternative Fuels Still Face Hurdles." AINOnline. September 1, 2010. Available at: http://www.ainonline.com [9/10/2010].

4 Ibid.

5 Coppa, Brian. "Military Biofuels Key to Rid US Foreign Oil Dependence," *Phoenix Green Business Examiner*, November 5, 2009. Available at: http: //www.examiner.com [9 September 2010]

6 Ibid.

7 Ibid.

8 Biello, David. "Navy Green: Military Investigates Biofuels to Power Its Ships and Planes," Scientific American, September 14, 2009.

9 Coppa, Brian. "Military Biofuels Key to Rid US Foreign Oil Dependence," *Phoenix Green Business Examiner*, November 5, 2009. Available at: http: //www.examiner.com [9 September 2010].

10 Biello, David. "Navy Green: Military Investigates Biofuels to Power its Ships and Planes," *Scientific American*, September 14, 2009.

11 Coppa, Brian. "Military Biofuels Key to Rid US Foreign Oil Dependence," *Phoenix Green Business Examiner*, November 5, 2009. Available at: http: //www.examiner.com [9 September 2010].

12 Biello, David. "Navy Green: Military Investigates Biofuels to Power its Ships and Planes," *Scientific American*, September 14, 2009.

13 Ibid.

14 Beidel, Eric. "Biofuels Industry: Friendlier Policies Needed to Displace Oil," *National Defense Magazine*, July 2, 2010.

15 "Military Signals Advanced Biofuels Demand: 336 million gallons by 2020," *BioFuels Digest.*

16 Epstein, Curt. "Alternative Fuels Still Face Hurdles," AINOnline. September 1, 2010. Available at: http://www.ainonline.com [9 September 2010].

17 Epstein, Curt. "Alternative Fuels Still Face Hurdles," AINOnline. September 1, 2010. Available at: http://www.ainonline.com [9 September 2010].

18 Epstein, Curt. "Alternative Fuels Still Face Hurdles," AINOnline. September 1, 2010. Available at: http://www.ainonline.com [9 September 2010].

19 Bradsher, Keith. "China Still Bans Rare Earth to Japan," *New York Times*, November 10, 2010; Broad, William J. "Mining the Seafloor for Rare Earth Minerals." *New York Times*, November 8, 2010.

20 Motavalli, Jim. "Forget Lithium—Its Rare Earth Minerals That Are in Short Supply for Electric Vehicles." June 19, 2010. Available at: http://www.bnet.com [10 September 2010]

21 Ibid.

22 Ibid.

23 Gordon, Julie. "Rare earth elements are used in a wide range of consumer products." *Reuters*. September 9, 2010.

24 Department of Energy, Critical Materials Strategy, December, 2010, p. 15.

25 Tu, Ninelu and Willi Teng. "Global Electric Car Market To Reach NT $3 trillion in 2014–2015." Digitimes, December 16, 2010.

26 "Toyota Promises 11 New or Redesigned Hybrids by 2013." Reuters. November 22, 2010.

27 Prior, Anna. "The Hidden Cost of Going Green." *Wall Street Journal*.

28 "Toyota Promises 11 New or Redesigned Hybrids by 2013." Reuters. November 22, 2010; Reed, John. "Appeal of the Electric Car 'Overhyped' Says Report." *Financial Times*, October 28, 2010, p. 18.

29 "Toyota Promises 11 New or Redesigned Hybrids by 2013." Reuters. November 22, 2010.

30 Reed, John. "Appeal of the Electric Car 'Overhyped' Says Report." *Financial Times*, October 28, 2010, p. 18; "Toyota Promises 11 New or Redesigned Hybrids by 2013." Reuters. November 22, 2010.
31 Ibid.
32 Ibid.
33 Moore, Heidi N. "Federal Money for Alternative Energy is Drying Up." *New York Times*, December 14, 2010.
34 "Renewable energy advocates push for extension of federal Treasury grant program." *Los Angeles Times*, December 8, 2010.
35 Gelsi, Steve. "GE aims for fresh renewable-energy deals." Marketwatch, December 15, 2010.
36 Robison, Peter and Gopal Ratnam. "Pentagon Loses Control of Bombs to China Metal Monopoly," Bloomberg, September 29, 2010.
37 Ibid.
38 Ibid.
39 Areddy, James T. and Nathan Hodge. "China Hold on Metals Worries Washington." *Wall Street Journal*, September 24, 2010, p. A19.
40 US Department of Energy, Critical Materials Strategy, December, 2010, p. 22.
41 Ibid, pp. 22–23.
42 Ibid, pp. 23–24.
43 Ibid, p. 24.
44 Ibid, p. 26.
45 Ibid, p. 26.
46 Ibid, p. 16.
47 Ibid, p. 24.
48 Gordon, Julie. "Rare earth elements are used in a wide range of consumer products." Reuters, September 9, 2010.
49 "China's Mineral Monopoly May Spur US Crisis." Chemical and Engineering News, September 2, 2010.
50 Ibid.
51 Bradsher, Keith. "After Rare Earth Embargo: A new Calculus for Toxic Work." *New York Times*, October 30, 2010, p. B1. Dombey, Daniel. "Washington Warns Against Dependence on China for Rare Earths." *Financial Times*, October 29, 2010, p. 6.
52 Bradsher, Keith. "After Rare Earth Embargo: A new Calculus for Toxic Work." *New York Times*, October 30, 2010, p. B1. Dombey, Daniel. "Washington Warns Against Dependence on China for Rare Earths." *Financial Times*, October 29, 2010, p. 6.
53 "US Rare Earths Concern is Well Placed to Profit." *Financial Times*, October 28, 2010, p. 18.
54 US Department of Energy, Critical Materials Strategy, December, 2010, p. 33–34.
55 Gordon, Julie. "Rare Earth Elements Are Used in a Wide Range of Consumer Products." Reuters, September 9, 2010.
56 "EU Looks for Alternative Sources," IBTimes.com; Gordon, Julie. "Rare Earth Elements Are Used in a Wide Range of Consumer Products." Reuters, September 9, 2010.

57 "US government worried about rare earth shortages," Hybrid Cars. December 15, 2010.

58 Ibid

59 Crooks, Ed. "US Rare-Earths Concern is Well Placed to Profit," *Financial Times*, October 28, 2010, p. 18; "EU Looks for Alternative Sources," IBTimes.com.

60 Fickling, David. "Playing Rare-Earth Roulette with Beijing," *Wall Street Journal*, October 5, 2010, p. C12.

61 Gordon, Julie. "Rare Earth Elements Are Used in a Wide Range of Consumer Products." Reuters, September 9, 2010.

62 Fickling, David. "Playing Rare-Earth Roulette with Beijing," *Wall Street Journal*, October 5, 2010, p. C12.

63 "China will need to import rare earths in 5 to 10 years." Hybrid Cars, October 5, 2010.

64 "Hybrid shortage: Australian rare earth project highlights problem." Hybrid Cars, October 1, 2010.

65 Bradsher, Keith. "China Still Bans Rare Earth to Japan." *New York Times*, November 10, 2010.

66 Crooks, Ed. "US Rare-Earths Concern is Well-Placed to Profit," *Financial Times*, October 28, 2010, p. 18.

67 Ibid.

68 Bradsher, Keith. "China Still Bans Rare Earth to Japan." *New York Times*, November 10, 2010.

69 Risen, James. "US Identifies Vast Mineral Riches in Afghanistan." *New York Times*, June 13, 2010.

70 Ibid.

71 Risen, James. "US Identifies Vast Mineral Riches in Afghanistan." *New York Times*, June 13, 2010; US Department of Defense Task Force for Business and Stability Operations in Afghanistan.

72 Risen, James. "US Identifies Vast Mineral Riches in Afghanistan." *New York Times*, June 13, 2010; Ricketts, Camille. "Afghanistan's Lithium Eureka: A Big Win for China, or Another Bolivia?" June 14, 2010. Accessed at http://www.venturebeat.com on 9/10/10.

73 Risen, James. "US Identifies Vast Mineral Riches in Afghanistan." *New York Times*, June 13, 2010.

74 Ibid.

75 Ibid.

76 US Department of Energy, Critical Materials Strategy, December, 2010, p. 31.

77 Ibid, p. 31.

78 Ibid, p. 32.

79 Ibid, p. 32.

80 Ibid, pp. 33–34.

81 Tabuchi, Hiroko. "China Urged to Resume Rare Earth Shipments." *New York Times*, October 25, 2010, p. B7; "China's Mineral Monopoly May Spur US Crisis," Chemical and Engineering News, September 12, 2010.

82 Ibid.

83 Hidge, Nathan and Stephen Difler. "Japan Scrambles for Rare Earth," *Wall Street Journal*, October 16–17, 2010, p. A10.

84 Areddy, James T. "China Cites Pollution in Tightening Rare-Earth Exports," *Wall Street Journal*, November 15, 2010, p. A 14.

85 "China's Mineral Monopoly May Spur US Crisis." September 2, 2010. Chemical and Engineering News.

86 Gordon, Julie. "Rare Earth Elements Are Used in Wide Range of Consumer Products." Reuters. September 9, 2010.

87 Robison, Peter and Gopal Ratnam. "Pentagon Loses Control of Bombs to China Metal Monopoly," Bloomberg, September 29, 2010.

88 Bradsher, Keith. "China Still Bans Rare Earth to Japan." *New York Times*, November 10, 2010; Tabuchi, Hiroko. "China Urged to Resume Rare Earth Shipments." *New York Times*, October 25, 2010, p. B7.

89 "China Resumes Rare Earth Exports to Japan," BBC News Business. Accessed at http://www.bbc.co.uk, November 29, 2010.

90 Bradsher, Keith. "China Still Bans Rare Earth to Japan." *New York Times*, November 10, 2010.

91 Tabuchi, Hiroko. "Block on Minerals Called Threat to Japan's Economy," *New York Times*, September 29, 2010, p. B1 and B4.

92 Bradsher, Keith. "China Still Ban Rare Earths for Japan, Executives Say." *New York Times*, November 11, 2010, p. B3.

93 Hodge, Nathan and James Areddy. "China Hold on Metal Worries Washington." *Wall Street Journal*, September 24, 2010, p. A 19.

94 "EU Looks for Alternative Sources of Rare Earth Minerals." IBTimes.com, November 26, 2010. Accessed at http://www.ibtimes.com.

95 Ibid.

96 Hodge, Nathan and Stephen Fidler. "Japan Scrambles for Rare Earth," *Wall Street Journal*, October 16–17, 2010, p. A10.

97 "China Resumes Rare Earth Exports to Japan." BBC News Business. November 29, 2010. Accessed at http://www.bbc.co.uk.

98 "Lynas and Sojitz to Distribute Rare Earths in Japan." November 27, 2010. Accessed at http://www.steelguru.com.

99 Bradsher, Keith. "In Dispute, China Blocks Rare Earth Exports to Japan." *New York Times*, September 23, 2010, p. B1 and B 12.

100 Dempsey, Judy. "EU Seeks Strategy to Reduce Reliance on China for Rare Earths." *New York Times*, November 25, 2010.

101 Broad, William J. "Mining the Seafloor for Rare Earth Minerals," *New York Times*, November 8, 2010.

102 Park, Sungwoo. "South Korea, Japan to Cooperate on Mining Rare Earths," Bloomberg, November 25, 2010.

103 Kate, Daniel Ten. "Japan to Build Nuclear Plants in Vietnam; Seek Rare-Earths Deal." Bloomberg Businessweek, October 31, 2010.

104 Bradsher, Keith. "China Still Bans Rare Earths to Japan." *New York Times*, November 10, 2010.

105 Hidge, Nathan and Stephen Fidler. "Japan Scrambles for Rare Earths." *Wall Street Journal*, October 16, 2010, p. A10.
106 "Japan develops rare earth free electric motor. " Hybrid Cars, September 30, 2010.
107 Ratnam, Gopal. "Pentagon Sees No Rare-Earths Crisis; May Aid US Producers." Bloomberg Businessweek, October 31, 2010.
108 Ibid.
109 Ibid.
110 US Department of Energy, Critical Materials Strategy, December, 2010, pp. 17–18.
111 Ibid, p. 40.
112 Ibid, p. 41.
113 Ibid, p. 42.
114 Ibid, p. 35.
115 Ibid, p. 36.
116 Ibid, p. 37.
117 Ibid, p. 35.
118 Ibid, p. 36.

References

Areddy, J.T. "China Cites Pollution in Tightening Rare-Earth Exports." *Wall Street Journal*, November 15, 2010, p. A 14.

Areddy, J.T. and N. Hodge. "China Hold on Metals Worries Washington." *Wall Street Journal*, September 24, 2010, p. A19.

Beidel, E. "Biofuels Industry: Friendlier Policies Needed to Displace Oil." *National Defense Magazine*, July 2, 2010.

Biello, D. "Navy Green: Military Investigates Biofuels to Power its Ships and Planes." *Scientific American*, September 14, 2009.

Bradsher, K. "After Rare Earth Embargo: A New Calculus for Toxic Work." *New York Times*, October 30, 2010, p. B1.

Bradsher, K. "China Still Bans Rare Earth to Japan." *New York Times*, November 10, 2010.

Bradsher, K. "In Dispute, China Blocks Rare Earth Exports to Japan," *New York Times*, September 23, 2010, p. B1 and B12.

Broad, W.J. "Mining the Seafloor for Rare Earth Minerals." *New York Times*, November 8, 2010.

"China's Mineral Monopoly May Spur US Crisis." *Chemical and Engineering News*, September 2, 2010.

"China Resumes Rare Earth Exports to Japan," *BBC News Business*. Available at: http://www.bbc.co.uk [November 29, 2010].

"China Will Need to Import Rare Earths in 5 to 10 years." *Hybrid Cars*, October 5, 2010.

Coppa, B. "Military Biofuels Key to Rid US Foreign Oil Dependence." *Phoenix Green Business Examiner*, November 5, 2009. Available at: http://www.examiner.com, on September 10, 2010.

Crooks, E. "US Rare-Earths Concern is Well-Placed to Profit." *Financial Times*, October 28, 2010, p. 18.

Dombey, D. "Washington Warns Against Dependence on China for Rare Earths." *Financial Times*, October 29, 2010, p. 6.

Epstein, C. "Alternative Fuels Still Face Hurdles." *AINOnline*. September 1, 2010. Available at: http://www.ainonline.com on September 10, 2010.

"EU Looks for Alternative Sources." *International Business Times*, November 26, 2010. Available at: http://www.ibtimes.com/articles/85993/20101126/europe-eu-australia-japan-china-minerals-rare-earth.htm.

Fickling, D. "Playing Rare-Earth Roulette with Beijing." *Wall Street Journal*, October 5, 2010, p. C12.

Gelsi, S. "GE Aims for Fresh Renewable-energy Deals." *Marketwatch*, December 15, 2010.

Gordon, J. "Rare Earth Elements Are Used in a Wide Range of Consumer Products." *Reuters*, September 9, 2010.

Dempsey, J. "EU Seeks Strategy to Reduce Reliance on China for Rare Earths." *New York Times*, November 25, 2010.

Hidge, N. and S. Difler. "Japan Scrambles for Rare Earth." *Wall Street Journal*, October 16–17, 2010, p. A10.

Hodge, N. and J. Areddy. "China Hold on Metal Worries Washington." *Wall Street Journal*, September 24, 2010, p. A19.

Hodge, N. and S. Fidler. "Japan Scrambles for Rare Earth." *Wall Street Journal*, October 16–17, 2010, p. A10.

"Hybrid Shortage: Australian Rare Earth Project Highlights Problem." *Hybrid Cars*, October 1, 2010.

"Japan Develops Rare Earth Free Electric Motor. " *Hybrid Cars*, September 30, 2010.

Kate, D. Ten. "Japan to Build Nuclear Plants in Vietnam; Seek Rare-Earths Deal." *Bloomberg Businessweek*, October 31, 2010.

Lane, Jim. "Military Signals Advanced Biofuels Demand: 336 Million Gallons by 2020."*BioFuels Digest*, December 16, 2010.

"Lynas and Sojitz to Distribute Rare Earths in Japan." November 27, 2010. Available at: http://www.steelguru.com.

Moore, H.N. "Federal Money for Alternative Energy is Drying Up." *New York Times*, December 14, 2010.

Motavalli, J. "Forget Lithium—Its Rare Earth Minerals That Are in Short Supply for Electric Vehicles." June 19, 2010. Available at: http://www.bnet.com [September 10, 2010].

Park, S. "South Korea, Japan to Cooperate on Mining Rare Earths." *Bloomberg*, November 25, 2010.

Prior, Anna. "The Hidden Cost of Going Green." *Wall Street Journal*, December 14, 2010. Available at: http://online.wsj.com/article/SB100014240527487044576045760117229537027 38.html.

Ratnam, G. "Pentagon Sees No Rare-Earths Crisis; May Aid US Producers." *Bloomberg Businessweek*, October 31, 2010.

Reed, J. "Appeal of the Electric Car 'Overhyped' Says Report." *Financial Times*, October 28, 2010, p. 18.

"Renewable Energy Advocates Push For Extension of Federal Treasury Grant Program." *Los Angeles Times*, December 8, 2010.

Ricketts, C. "Afghanistan's Lithium Eureka: A Big Win for China, or Another Bolivia?" June 14, 2010. Available at: http://www.venturebeat.com on September 10, 2010.

Risen, J. "US Identifies Vast Mineral Riches in Afghanistan." *New York Times*, June 13, 2010.

Robison, P. and G. Ratnam. "Pentagon Loses Control of Bombs to China Metal Monopoly." *Bloomberg*, September 29, 2010.

Tabuchi, H. "Block on Minerals Called Threat to Japan's Economy." *New York Times*, September 29, 2010, p. B1 and B4.

Tabuchi, H. "China Urged to Resume Rare Earth Shipments." *New York Times*, October 25, 2010, p. B7.

"Toyota Promises 11 New or Redesigned Hybrids by 2013." *Reuters*. November 22, 2010.

Tu, N. and W. Teng. "Global Electric Car Market to Reach NT $3 Trillion in 2014–2015." *Digitimes*, December 16, 2010.

US Department of Energy, *Critical Materials Strategy*, December, 2010.

"US Rare Earths Concern is Well Placed to Profit." *Financial Times*, October 28, 2010, p. 18.

Chapter 6

Hydropolitics in South American International Relations: A Perspective on Water Governance at the Prata Basin – Between Conflict and Cooperation (1960–1992)

Fábio Albergaria de Queiroz

Introduction

The planet has about 1,386 billion km³ of water. Almost all of it – 97.5 percent – is salty. Of the 2.5 percent of freshwater, more than two thirds are not available for consumption or industrial uses, since most of it is contained in glaciers or in permanently frozen ground (permafrost) (Clarke and King, 2005: p. 20). Only with high difficulty, and high costs, the available amount of this resource can be increased. Thus, freshwater "technically" available for human use is only a very small proportion usually found in easily accessible areas such as lakes, rivers and wetlands (Siranneau, 1996; Tundisi, 2003):

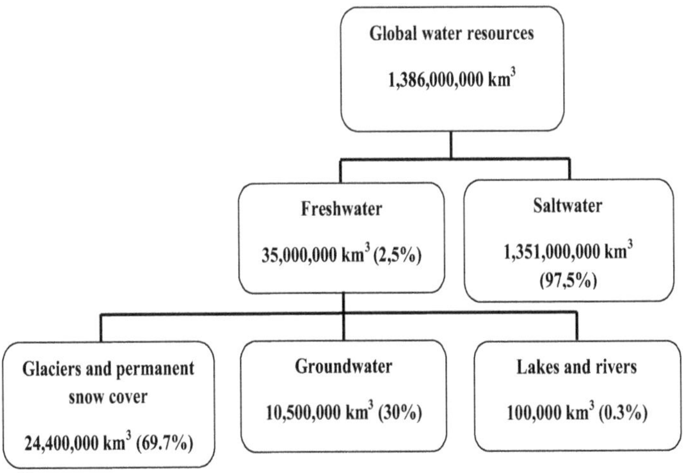

Figure 6.1 Distribution of Global Water Resources
Source: Clarke and King (2005: p. 20).

Figure 6.2 International River Basins of South America

Source: Available at: http://www.transboundarywaters.orst.edu/publications/register/images/southamerica.gif.

Moreover, there is high level of inequality in the geographical distribution of freshwater as its availability varies in different regions of the world. This scenario invites us to reflect on water-related issues as a factor capable of generating instability. After all, to respond to some multiple internal requirements of a country, from human needs to power generation, using water that does not originate solely within national borders, but from resources shared with other states, proves to be a strategic issue in the field of International Relations.

In relation to shared rivers, a relevant fact is that only a quarter of them have their use regulated by effective international treaties, and more than half of them do not provide any mechanisms for dispute resolution (Ferreira, 2003: p. 39). In this complex context in which water resources are shown as a reference point in International Relations, South America stands out because its 38 shared basins represent one of the largest reserves of freshwater in the world, approximately 26 percent of the total available (see Figure 6.2).

On the other hand, the arid and semi-arid regions of the Middle East, Sub-Saharan Africa and Southeast Asia concentrate the most critical water stress levels (water availability per capita between 1,000 and 1,600 m^3/year) or absolute scarcity of water (less than 500 m^3/year) (UNESCO, 2003: p. 9). Based on these considerations, and for purposes of illustration, if the world map was conceived from the perspective of freshwater distribution, we could have the format below, highlighting South America as a region that clearly stands out for concentrating most part of the total amount available:

In reviewing this framework and taking into account the relationship between water versus population, there is mounting evidence that the distributional inequalities are increasing, especially in higher population density regions. This is the case, for example, with the Asian continent which has about 36 percent of the total available water resources, but needs to meet the demands of more than half of the world population, as shown in the following figure:

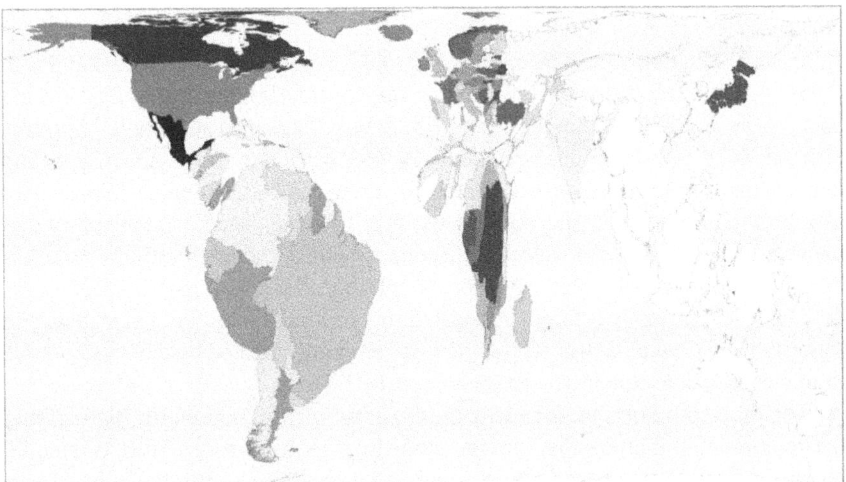

Figure 6.3 World Map from the Perspective of Distribution of Water Resources

Source: www.worldmapper.org.

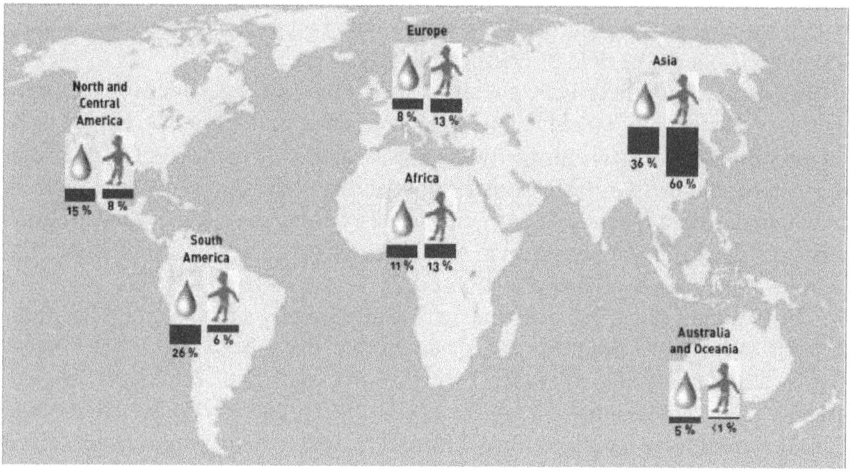

Figure 6.4 Water Availability Versus Population
Source: UNESCO (2003: p. 69).

Hence, shared waters have the potential of generating highly complex relations, as the actions of one country can undermine the interests of its neighbors. This is what Elhance (1999: p. 13) labeled *hydropolitical interdependence*. This situation underscores the importance of territoriality and therefore, the level of regional analysis as the most appropriate form to deal with issues of hydropolitics[1] and security that arise among those who are the main actors in this inter-relationship: the states (Buzan et al., 1998: p. 199).[2]

In such situations, the concept of vulnerability described by Keohane and Nye (1989: p. 12) as the responsiveness of a state, considering the availability and the costs of the alternatives that these actors may have, becomes an equally important element in the international relations of countries that share the usage and management of water resources. Water, for these reasons, is regularly presented as a protagonist of dramatic predictions. Due to issues such as availability and uneven distribution, the prospect that social, political and even military conflicts over the allocation of water will increasingly become a key part of the twenty-first century landscape is frequently stated by prominent international leaders and scholars (Van Ginkel, 2001). Thus, water is commonly treated by them as a source of power and, consequently, a national security priority.

Therefore, so many actors underscore the sensitivity of this issue by highlighting the multifaceted matters that emerge from hydropolitics as possible existential threats, especially in those regions where water scarcity and/or the existence of a high interdependence resulting from the shared uses of water resources stand out as an important feature in the relations of the riparian states (Hinrichsen et al. 1998; Turton, 2001; Allan, 2001; Yetim, 2003; Clarke and King, 2005: p. 75; Procópio, 2007: pp. 271–3; Kistin, 2007).

In such a scenario, we are led to reflect on the possibility of situations of high tension being motivated, in the near future, by the competition for water resources. This is, for instance, the standpoint defended by Ismail Serageldin, former World Bank vice-president, who asserted that if the wars of this century were fought over oil, the wars of the next century will be fought over water (Le Prestre, 2000: p. 443). In the same direction, the British government warned that there is a growing likelihood of wars over water resources in a not so distant future. Authorities have also announced that British armed forces "must be prepared to intervene in humanitarian disaster relief, peacekeeping and warfare related to dwindling natural resources, particularly water" (*The Independent*, 2006 quoted in Castro, 2007: p. 108).

In line with the foregoing context, Hinrichsen et al. (1998) see the water issue, particularly in relation to its scarcity, as a factor of instability in international relations, after all, complement Clarke and King (2005: p. 19), currently over one billion people lack access to fresh water and estimates indicate that in 2050 more than half of the projected world population will live in countries with chronic shortage of this resource.

These statements and admonitions are not entirely surprising given that over the last few decades international security experts have warned that water was becoming a potential source of conflicts around the world (Gleick, 1993; Gleick, 2000), especially considering that 263 river basins, where almost half the population of the world inhabits, are shared by two or more countries (Cosgrove, 2003: p. 1). It is also estimated that fewer than ten countries control about 60 percent of the freshwater sources available, and a large number of groundwater aquifers are shared by two or more countries (Ohlsson, 1995; Castro, 2007: p. 12).

On the other hand, despite the arguments presented, another relevant group of scholars asserts that the hypothesis of a hydropolitical global scenario marked by instability and increasing tensions around water is based on premature statements often exaggerated by biased assessments, as noticeable in many speeches, and even in official documents of specialized agencies. This group contends that there is little historical evidence to support the conflictive hypothesis raised. In contrast, they argue that peaceful cooperation in sharing water resources has historically been the default behavior and the use of violence, therefore, the exception (Libiszewski, 1995; Wolf, 1998; Salman and Boisson de Chazournes, 1998; Allan, 2001; Cosgrove, 2003: pp. 10–11; Yoffe et al. 2004).

In addition, Clarke and King (2005: p. 75) highlight that over the past few years many countries have created management committees to deal with the common usage of important rivers like the Rhine, the Zambezi, the Nile, the Mekong and the Danube. Europe, for instance, has almost two hundred treaties regulating the use of such waters. Therefore, although the relevance of the admonitions concerning the risks of water conflicts, Le Prestre (2000: pp. 427–8) warns about the danger of reductionism when dealing with this issue. According to him, in situations where the uses of water are subject to dispute, it is always possible to build a direct or indirect bond, close or distant, between the instability to be explained and a hydropolitical variable and thus, see as a result of this dangerous combination, the announcement of future instabilities.

Summing up, the arguments and facts presented suggest caution when dealing with the binomial relation water-security, demonstrating the need to measure the real influence of water on the multitude of other variables capable of exacerbating a situation of instability or problem resolution. Disagreements over water usually take place on countries where there is a history of animosity and resentment, however, it must be pointed out, not always water will be the pivot for disputes that might arise. Generally, water constitutes an aggravating factor of a pre-existing framework of tension catalyzed by a scenario of high hydrological interdependence.

Hence, it becomes more useful and feasible to study hydropolitics as one of the factors to be considered in security policies, a factor capable of generating additional stress in conflict situations, and not as the primary cause of their outbreak, though on occasion, it is presented as the latter. As a result, bringing the arguments presented together, we can summarize the characteristics that shape the discourse at the hydropolitical arena using the synthesis made by Du Plessis (2001: pp. 13–14) (quoted in Turton, 2003: pp. 74–5), introducing it as:

> [...] concerned with, and intrinsically linked to, the concept of security. This concern extends to environmental security in general, and to water security in particular. This latter focus, and its collateral theoretical conceptualizations, is forced upon the scene by specifically linking the water discourse ... to the war/peace and to conflict/cooperation [problematique] and by considering water to be a potential source or cause of (violent) conflict. [...] This development is, in part, the result of the new security paradigm that has broadened and deepened the security agenda by including non-military (low politics) threats, as well as non-state, security stakeholders at all levels of society. Hence, it is also linked to the notion of common security, which has as its foundation common interests that, at a minimum, requires a shared interest in survival. Irrespective of the fact that post-1989 security has acquired a wider meaning than protection from military threat, its broader conceptualization has paradoxically contributed to securitization and militarization of water as a traditional non-military concern.

Considering this brief introduction, we may assert that water can aggravate conflictive situations, in terms of:

1. the degree of vulnerability and dependence of each country in relation to its multiple uses,
2. the number of actors claiming access to it,
3. power resources of these actors,
4. the existence of substitute products (non-existent at present) or, even,

5. the symbolic character water assumes in political, cultural and identity issues.

Therefore, the scenario described above allows us to infer that the multidimensional issues that emerge from hydropolitics have the potential to affect, in a greater or lesser extent, depending on the intervening variables considered, the political balance between states and other actors that share the waters of a hydrographic system. Consequently, hydropolitics emerges as an important factor to be considered in setting regional geopolitics and, hence, the composition of Regional Security Complexes agendas.

Based on these considerations, the central question to be addressed over the next few pages is: in the context of regional hydropolitics, how could the Prata Basin, one of the major hydrological systems of the planet, influence the shaping process of international relations in the South American Regional Security Complex? The problem raises a second question: Could the dynamics of hydropolitics in the South American context become so intense as to make the shared use of the Prata Basin resources a security issue? If so, why?

In seeking answers to these questions, this study assumes the following hypothesis: the greater the degree of hydrological interdependence between actors that share the uses of a water system, the greater the probability of formation of Hydropolitical Complexes (HC). Therefore, this descriptive model, an adaptation developed along the lines of the Regional Security Complexes Theory to deal with specific relationships that may potentially arise regarding the common use of water resources, shall be used as a tool in this study.

Thus, in order to meet the objectives proposed, we will first outline this expanded conceptual framework. Afterwards, we will attempt to verify whether in the case of the South American Regional Security Complex there is confirmation of the hypothesis raised amongst the actors that share the water resources of the Prata Basin.

An Overview of the Expanded Concept of Security

The discussion over the enlargement of the security concept is rooted in the contributions made by the School of Copenhagen that gives us a symbolic reference point. The methodology adopted by this group gives to researchers an alternative theoretical approach to the traditional one.[3] So, given this broader perspective, how do we now define what security really is?

For Barry Buzan, Ole Wæver and their collaborators, security is "the move that takes politics beyond the established rules of the game and frames the issue either as a special kind of politics or as above politics" (Buzan et al. 1998: p. 23). This process, which they called securitization, may be understood as an extreme version of politicization in which the securitizing actor facing a threatening situation tries to put the object to be protected (referent object) in a locus of decision immune to the ordinary rules of the political scenario. As a result, they may use the means they feel are necessary to solve the problem, including the force.

It should be noted that a given issue can be framed in a broad range of ways, from *non-politicized* (where it is not seen as a topic that should be submitted to public debate or demand decisions in terms of security) to *politicized* (where there is the recognition that the subject is of public interest and therefore, subject to governmental decision and resource allocation) and, in extreme cases, as *securitized* (as mentioned earlier, a condition which demands the use of emergency measures, even outside the normal bounds of political procedure).There is also the inverse movement, *desecuritization*, which consists of moving subjects previously classified as existentially threatened to the level of the usual bargaining and dispute of the political sphere (Buzan et al. 1998: pp. 23–4; Sheenan, 2005: p. 54). It should also be noted that typifying a matter in accordance with these conceptual structures is not an exclusive prerogative of states and may also occur through the agency of non-state actors.

This demonstrates, then, that the accuracy of the criteria established to define certain topics such as security is an intersubjective social construction, and is directly related to the discursive emphasis given to possible substantial political effects of a given existential threat to a referent object, which reveals the central role of speech-acts in this approach.

Speech-Act and Securitization

What is the nature of the relationships established between words and the real world? This question, raised by several scholars of the field, is the basis of the Theory of Speech-Acts (TSA), a perspective of the Analytical Philosophy of the Language that posits that this system of signs used in the communication of ideas, the language, is a concrete social practice for the execution of actions, a form of constitution of reality and therefore an inseparable part of the real world that gives form and meaning to it in a complex relationship of interdependence (Searle, 1984; Austin, 1990).

From this premise, the TSA believes that language should not be merely considered in its simple abstract form (locutionary form), but always within a social and cultural context in which the linguistic expressions and sentences handed down by the speakers make sense in relation to components of context (illocutionary act) producing, then, effects and consequences (perlocutionary acts).[4]

Constructivist Theory – the theoretical basis of this analysis – after assimilating those basic assumptions, started to emphasize the role of discursive processes as they directly reflect the ideas which are relevant to the actors and, therefore, their interests. In this context, Wendt (1999: p. 135) postulates that when material variables are investigated, equal attention should be devoted to analysis of the discourses underlying them, as this interrelation demonstrates the validity of the arguments raised by a given actor and thus the potential of their material forces.

Therefore, constructivists view language as a vehicle for dissemination and institutionalization of ideas, a mechanism for the construction of social reality (Adler, 2006: p. 103) as does Language Theory. Thus, the TSA labels the speech-act not only as a describer of the reality, but also as a tool to shape the structure through which things "mean" and, at the same time, it has meaning given by the authorized speakers in which they impart qualities, attributes and interrelationships with other objects (Kratochwil, 1989: p. 8; Milliken, 2001: pp. 138–40).

As theorized by the School of Copenhagen, these assertions can be applied in the case of security studies once speech-act is a central element for their proper understanding because, as explained by Wæver (1994: p. 5), "we can regard security as a speech-act". Thus, a subject assumes the status of securitized not necessarily because of the existence of a real threat, but because it is presented as such by the speech-act in a given context. Therefore, by using the speech-act to this end, this actor posits in his favor the right to use extraordinary means to counter the threat, whatever they may be, even if it means breaking the normal rules of the political game. In this regard, as stressed by Woods (1996: p. 27), the speech is synonymous of an action, which means to say that it signals, in fact, practice and behavior. Furthermore, as Sheenan (2005: p. 55) asserts, "securitizing is never an innocent act".

Table 6.1 Application of the Theoretical Model of the School of Copenhagen to Hydropolitics

	According to the Copenhagen School	On Hydropolitics
Referent Objects	Things that are seen to be existentially threatened that have a legitimate claim to survival.	Political, economical, societal, territorial and/or environmental stability of a state on account of the actions of some riverside in a particular water course.
Securitizing Actors	Actors who securitize issues by declaring a referent object as being existentially threatened.	States which, through their specialized bureaucracies and sub-agents representing the national decision-making elite, externalize their perceptions of the threats as a result of sectoral actions in a certain water course.
Functional Actors	Actors who affect the dynamics of a sector without being the referent object or securitizing actor.	Third parties that have some kind of interest in the outcome of the situation as: riparian states little or not affected, International Organizations, Non-governmental organizations, pressure groups, epistemic communities, riparian communities.

Source: Turton (2003: p. 79).

However, this process has limitations. A speech that presents something as an existential threat regarding a referent object does not necessarily establish a situation of securitization. To accomplish that, the securitizing actor must convince his audience that the adoption of special measures is needed to counter the threat and ensure the safety of the object. Thus, securitization occurs "only if and when the audience accepts it as such" (Buzan et al. 1998: p. 25).

Such a process, labeled by the authors as a movement of securitization, is configured as a negotiation between a securitizing agent and its audience by which this agent will seek legitimacy and permission to disregard the existing rules of the political game and then change them. Even if there are signs of such acceptance by the audience this does not constitute the securitization of the object, but simply a movement of securitization.

This analysis supports the premise advocated by Buzan (1991: p. 60) that the state is not a coherent and unitary actor in terms of perceived threats to its integrity, although, despite the divergences, there is always a result, even temporarily, of the alignments and domestic political coalitions (Viola, 2005: p. 8). Therefore, unlike the neo-realism perception of threat "as something almost material and objective" (Stahl, 2007: p. 4), the group of Copenhagen presents threats, following the same logic of the definition of security, as a social construction.

The Security Complex Theory

In addition to proposing the expansion of the concept of security, the School of Copenhagen has brought another great contribution to the area, highlighting the importance of the dynamics that occur at an intermediate level of analysis. Because of this, they've suggested, as a conceptual reference, the descriptive model of Regional Security Complexes (RSC). The basic premise of this model assumes that in the post-Cold War era, international relations in the security field tend to be regionalized since the collapse of the bipolar world order has reduced the incentives that led Washington and Moscow to globalize the systemic agenda of security, which, therefore, led to greater autonomy and space for the local dynamics and actors (Buzan and Wæver, 2003).

Moreover, geographical proximity proves to be a relevant variable now that sector threats move more easily between shorter distances. Therefore, as indicated by Buzan and Wæver (2003: p. 27), the security relationship from the subsystem level of analysis, located between the global and local, is what gives great analytical power to the Theory of Regional Security Complexes.[5] As presented by Barry Buzan in *People, States and Fear: An Agenda for International Security in the Post-Cold War Era* (1991: p. 198), the concept of RSC was stated as "*a set of states* whose main perceptions and security concerns are so interlinked that their national security problems can not reasonably be analyzed or resolved separately". The undeniable state-centrism present in this concept is part of its ontological composition, so that only states are considered as units within these small sub-anarchic systems, restricted to the military and political sectors.

However, with sector expansion and the introduction of new actors, the revision of the original concept of RSC has become urgent in order to extend its analytical capacity and thus bring it to this new theoretical and conceptual structure. To fulfill this purpose, Buzan et al. (1998: p. 201) redefined it as "*a set of units* whose processes of securitization, dessecuritization, or both, are so interlinked that their security problems cannot reasonably be analyzed or resolved apart from one another".

Still, even with the incorporation of new issues and actors, one can see that the RSC are designed considering the state as the basic unit of analysis, because they are presented in all sectors as central actors and main interlecutionists. This regularity does not occur with other ontological unities, for example, the nation that is important in the societal sector, but not necessarily in the environmental and/or economic.[6]

We see, therefore, that the RSC is a result of symbiotic interactions between material and ideational variables such as beliefs, identities, material capabilities, borders, distribution/perception of power, anarchy, polarity and social construction, which gives the model greater scope, consistency, and analytical/ontological density. Another central point of this structure, one of the distinguishing features of a RSC, is the fact that the processes of securitization/desecuritization and, therefore the degree of interdependence in security, is markedly more intense between the component units than between them and actors outside the RSC (Buzan and Wæver, 2003: p. 4). At the same time, if the interests of one or more exogenous powers exert influence to the point of removing the internal dynamics of the Complex, we have what Buzan et al. (1998: p. 12) labeled as overlay, a situation that hinder the formation of the RSC. So, when this type of overlap occurs, it is not possible to identify a pattern from the local dynamics and, consequently, the existence of a Complex derivate from these relations, because the interests of external powers define the security agenda in the area overlain (Buzan et al. 1998: p. 12).

From this general perspective, the authors outlined two possible configurations where these new actors and sectors are incorporated:

1. Homogeneous Complexes – are those composed of specific forms of interaction between similar units (e.g., military complexes formed by states);
2. Heterogeneous Complexes – are those that combine different types of actors in two or more sectors (for instance: states, nations and transnational companies interacting across the political, economic and societal sectors).

Identified this strong introspective characteristic of the relationships that define a RSC, another variable stands out closely related, and equally important in this structure: the relations of amity and enmity between their units from the interactive processes of identity formation. According to the standards of amity –

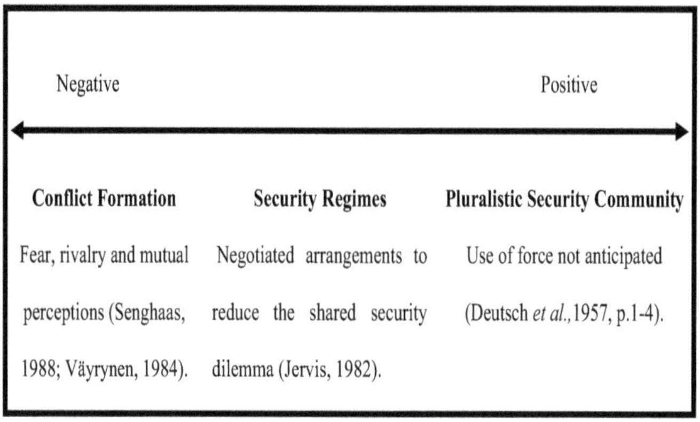

Figure 6.5 Classification of RSC from the Patterns of Amity/Enmity
Source: Turton (2003: p. 76).

enmity, shaped by historical and social dynamics, the RSC can be classified as a continuum with a positive pole and a negative one.

In the positive extreme the friendly relations between units are preponderant and give rise to the formation of a Security Community where the expectations of change are always peaceful (Deutsch et al. 1957).[7] On the other hand, in the negative pole, these relations are tensioned by mistrust and fear and result in conflicting formations (Väyrynen, 1984; Senghaas, 1988). In the intermediate zone between these two poles it is possible to observe the formation of security regimes (Jervis, 1982), a hybrid configuration, as schematically presented above (Figure 6.5).

Based on the propositions of the School of Copenhagen, we can summarize the structure of a RSC from three main types of relations:

1. the distribution of power between the units in the region (polarity), which requires an anarchic structure composed of two or more autonomous units;
2. the patterns of amity/enmity between the actors in the military, political, societal, economic and/or environmental sectors;
3. the power relations with actors outside the region, especially with respect to the superpowers and the major powers.

Thus, from this set of variables – anarchic structure, polarity, social construction and boundaries – Buzan and Wæver (2003: pp. 445–6) identified the existence of eleven RSC in the post-Cold War: North America, South America, Europe, Post-Soviet Union, Middle East, West Africa, Central Africa, the Horn of Africa, Southern Africa, South Asia and East Asia.

According to this typology, South America has usually been classified as a RSC of the standard type, marked by the presence of more than one regional power.[8] Moreover, it is important to note that the RSC is composed of two sub-regional structures: the Southern Cone Subcomplex, formed by Argentina, Bolivia, Brazil, Chile, Paraguay and Uruguay, and the Andean Subcomplex bringing together Peru, Ecuador, Colombia, Venezuela and Guyana.[9]

Hydropolitics and the Theory of Regional Security Complexes

A panoramic look over the scenarios of international hydropolitics signals us that there are regions where water resources emerge, under the terms of the School of Copenhagen, as an issue clearly securitized. The shared use of resources of a hydrological structure, for various reasons, has the potential to influence international relations among actors who depend on them, showing further that the perceptions and concerns that emerge from hydropolitics may be linked to such an extent to national security problems that they cannot reasonably be analyzed or resolved apart from one another.

Thus, in an attempt to search for evidence on possible connections between the dynamics originated in the specific context of the Tigris-Euphrates basin and the security agenda of the RSC of the Middle East, Schulz (1995: p. 97), using the Regional Security Complexes Theory, formulated the concept of Hydropolitical Security Complex (HSC). According to the author, this is a special form of Security Complex that arises when "a set of states that are geographically part owners and users of technically shared rivers start to consider, as a consequence, this water body to be a major national security issue".

It should be also noted that this scenario can get more unstable if there is a history of enmity between countries in the Complex and/or, as added by Bachler (1995, quoted in Ferreira, 2003: p. 50), when the mechanisms of integration, cooperation and disputes resolution between riparians are ineffective or non-existent increasing the possibility of using extraordinary means, including force, to cope with existing demands. In this peculiar context Schulz (1995: p. 97) identified the existence of an HSC formed by Turkey, Iraq and Syria.

This theoretical construction was based on the identification of an intrinsic interdependence between the national security agendas of each country derived from cross-border effects caused by the use of shared waters of the Tigris-Euphrates basin in relation to the construction of dams, water flow, salinization and hydroelectric power generation.

The condition of the Tigris-Euphrates Hydropolitical Security Complex, as labeled by Schulz, became even more intense because the region is characterized by severe drought (Buzan et al. 1998: p. 90) exacerbating the perception that the regional hydropolitics was, in this case, a matter of strategic priority. From the analysis undertaken, Schulz has systematized what he described as vertical and horizontal relationships. While vertical relations refer to the possible connections between the Hydropolitical Security Complex and the global systemic level, for example, by the intrusion of exogenous powers in the affairs of the Complex, the horizontal relations refer to the dynamics that occur in the same structural level, i.e., within and between Complexes.

In short, the work of Schulz identifies, through this conceptual framework, the dynamics that take shape in the field of hydropolitics where the management of shared water resources is, in some way, associated with national security of the actors in an HSC, or in other matters of high politics capable of connecting them.

Despite its importance, this model has limitations as to its applicability in situations that do not exhibit the same density of conflict, as in the case analyzed by Schulz. Thus, the HSC, as originally proposed, takes shape in stressful situations, scenarios prone to securitization of water in its more negative spectrum, i.e., considering the use of extraordinary means. Furthermore, the model was developed focusing on the sharing of surface water. Yet, there are some important points for reflection. First, is it possible to have similar structures in a situation of hydrological interdependence, but in a context of cooperation? Moreover, can the dynamics arising from the sharing of groundwater and aquifers, large reserves of freshwater, also occasion the formation of Hydropolitical Complexes?

Taking these considerations and having as reference the work of Schulz and the contributions of the School of Copenhagen, authors such as Allan (2001) and Turton (2003) improved the original model by incorporating these new dimensions in order to give it greater explanatory power. To this end, they have introduced the constructivist premise of amity/enmity in a wider continuum, as they defined the axis of the dynamics of security. They also gave greater attention to the sharing of underground water systems, and not just the superficial ones, as a reference to explain the role of hydropolitics in the political agenda of sub-regions such as West and South Africa and the Middle-East.

From these empirical tests, Allan (2001, 2002) and Turton (2003, 2008) began to contemplate the existence of what they called Hydropolitical Complexes in regions where the dependence on shared water systems – both surface and groundwater – is of such a strategic nature that this dependence starts to drive inter-state relations of potential cooperation (amity) and/or competition (enmity) in a discernable manner (Turton, 2008: p. 188).

Therefore, the authors have agreed that Hydropolitical Complexes (HC) normally occur in situations where relations between riparians tend to strengthen the bonds of amity and mutual cooperation. On the other hand, if these dynamics indicate a pattern of hostility and tension, as in the case involving Turkey, Iraq and Syria, we have, as previously defined by Schulz, the possible formation of a Hydropolitical Security Complex (HSC). So, according to this typology, the HSC turned out to be classified as a distinct type of Hydropolitical Complex.

Thus, the model of the HC/HSC, originally presented as an integral part of a heterogeneous Regional Security Complex (Turton, 2001: p. 19), makes it possible to verify both the existence of multi-sectoral cleavages (in cases of horizontal relationships) and connections between different levels of analysis (vertical relations) in scenarios of high hydrological interdependence, marked by a continuum of cooperation and/or conflict.

Hydropolitics in the Prata Basin: Between Conflict and Cooperation (1960–1992)

Historically, the Prata Basin has proved to be a region highly sensitive not only to the countries that comprise it, but to the balance of power in South America. International Relations in the region experienced, throughout history, alternating periods of antagonism, conflict, mistrust and cooperation. The Prata Basin, with an area of 3.1 million km² (equivalent to 17 percent of the total area of South America), is the second largest hydrographic basin in the continent, only surpassed by the Amazon Basin. This hydrographic system covers the territories of five countries: Argentina, Bolivia, Brazil, Paraguay and Uruguay.

Most of the basin area is located in Brazil, totaling 1,415,000 km², followed by Argentina (920,000 km²), Paraguay (410,000 km²), Bolivia (205,000 km²) and Uruguay (150,000 km²). The main water course of the basin is the Paraná River, with a length of 3,780 km, together with its major tributaries: the Paraguay, with 2,620 km, and Uruguay, with 1,600 km. The Parana becomes the Prata River when it receives the Uruguay River, just before flowing into the Atlantic Ocean:

The economic importance of the basin is uncontested since 70 percent of the five countries' GDP is generated in its domains, and about 50 percent of their combined populations reside therein (Brazilian National Water Agency, 2001: p. 5). Its hydroelectric potential is 90 billion kWh per year and its main rivers provide 15 thousand km of navigable waterways, being the only natural exit to the sea for Bolivia and Paraguay, as well as to important regions of Brazil and Argentina (Schilling; 1981: p. 122; Pinto, 2008: p. 18).

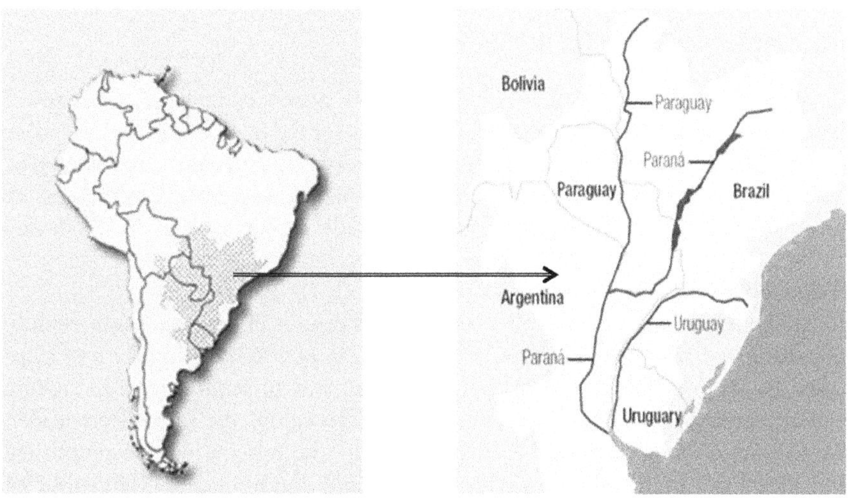

Figure 6.6 The Prata Basin

The construction of the Binational Power Plant of Itaipu, between Brazil and Paraguay, made the region's energy potentional one of the most important issues on the international relations agenda of the Southern Cone.[10] This enterprise has brought entanglements in different fields, involving not only its members but also Argentina, a reflection of past problems and unresolved geopolitical regional challenges immersed in distrust.

In relation to Paraguay, Itaipu brought up issues involving borders that have been dormant since the end of the war against the Triple Alliance formed by Brazil, Argentina and Uruguay (1864–1870).[11] While Argentines saw in this partnership the expansion of Brazilian geopolitical influence in the Southern Cone, they also saw the reduction of theirs (Pecequilo and Hage, 2007: p. 2). Even today, Itaipu illustrates a complex interdependence with the potential to generate risk and vulnerability of economical, political, social and environmental dimensions to the region, demonstrating therefore to be a matter that requires attention.

Furthermore, the Paraguay-Parana waterway – a project of great magnitude started in the 1980s to integrate the five countries along its 3,440 km confirmed a change of perception in the international relations of the Southern Cone marked, especially, by multilateralism and the pursuit of cooperative arrangements. For this reason, it is observed that the relevance of the South American hydropolitics points to singular issues that grant to the integrated management of water resources theme, a strategic role in the international relations of the region. This situation demands a regulatory structure capable of establishing rules that institutionalize common responsibilities and duties, once most of the water resources are shared. However, the formation of such an institutional apparatus is a complex task, because it may conflict with the sovereign interests of each of the riparian states, as we shall see.

Hydropolitical Dynamics in the Prata Basin

In early 1960, driven by the expansion of its economy and the concomitant increase in energy demand, Brazil began studies for the hydroelectric exploitation of the Parana River, in the contiguous stretch between the country and Paraguay. Moreover, the prevailing hostilities between Brazil and Argentina as well as the incompatibility of its hydroelectric projects would end up scoring the 1960s and 1970s as a predominantly realist phase of international relations in the Prata Basin (Filho, 2006).

It should be noted that prior events in this region in the nineteenth century foreshadowed what was to come in future years. In 1863, one of the first steps taken by the Paraguayan leader, Solano Lopez, was to send a troop to explore and demarcate the region. Arriving at his destination, the expedition leader, Lt. Domingo Patiño, recorded in his diary that "the mission was accomplished [...] then I put in the place a wooden plaque with the initials R.P. (Republic of Paraguay)" (quoted in Pinto, 2008: p. 35). The problem was that the region, for the Paraguayans known as *Salto de Guaira*, in Brazil *Sete Quedas* (Seven Falls), was held as part of the territory of both countries, a problem that remained inert until the 1960s despite of its apparent centrality to the balance of power in South America.[12]

Moreover, despite the problem of borders, the Brazilian initiative to make exclusive use of the river, ignoring the communal waters regime, seemed to undermine the sovereignty of Paraguay (Pereira; 1974: p. 51). One of the first measures taken by Brazilian President João Goulart, trying to address the major energy crisis that was about to explode, was to propose to the Paraguayan President, Alfredo Stroessner, a joint plan for hydroelectric development of the Paraná River, an initiative that took place on 19 January 1964, just over two months before Goulart's overthrow by the military coup.

However, with the early termination of the mandate of Goulart and the arrival of the military forces into power on 31 March 1964, the course of the negotiations changed. The first step in dealing with this situation of uncertainty and mutual distrust was taken by the Castello Branco administration (1964–67), the first of the military regimes in Brazil (1964–1985), which had as one of its goals to make the country the dominant power in the Southern Cone region. For this, it was necessary to enhance economic growth and energy generation since, as mentioned, it was essential that these plans succeed. But how could they do it? How could they overcome the mistrust and territorial disputes of the past?

The nomination of the engineer Otávio Marcondes Ferraz as president of Eletrobras, the agency responsible for the Brazilian energy sector, polarized the debate on this subject in the country. He was opposed to the idea of building a binational power plant in Sete Quedas and advocated the diversion of the Parana River, so that the use of its waters could be made exclusively in Brazilian territory (Filho, 2006).

The theme, as might have been expected, continued to cause problems for Brazilian diplomacy until, in 1966, after much discussion, the Brazilian chancellor, Juracy Magalhães, expressed, to the Paraguayan government, Brazil's intention to initiate a joint Project for a future hydroelectric exploitation of *Saltos de Guaira* (*Sete Quedas*), which was considered in Asunción as a considerable progress to resolve the differences.

All this diplomatic engineering, which lasted for four years, from 1962 to 1966, was accompanied by great concerns on the part of government of Argentina, a downstream country with major interests on the hydro-electric potential of the Parana River. Thus, whether an exclusively Brazilian project or binational, with or without diversion of water, with this project consolidated, Argentina would be vulnerable to claims of its biggest rival, Brazil. Thus, in an attempt to narrow Brazilian projection in the Prata Basin and to turn Argentina into an alternative pole, the President Arturo Illia (1963–1966) proposed, on the pretext of seeking the regional development, the outline of a plan for the physical integration of the basin based on the exploitation of its natural resources.

In short, in the geopolitics of South America from this period, the states perceived each other as potential threats, however, they have sought, albeit precariously, to avoid an escalation of conflict between them through cooperation, appropriately defined by Keohane (2005: p. x) “as an intensely political process of mutual adjustment in a situation of actual and potential discord”. After all, as said by Waltz (1959: p. 182), “in anarchy there is no automatic harmony”.

The Act of Iguaçu and the Construction of a Hydropolitical Regime in the Prata Basin (1966–1970)

In 1966 it was signed the Iguaçu Act (or Cataratas Act), an initiative considered as the initial milestone that later would lead to the Treaty of Itaipu, the architect of the largest hydroelectric plant on Earth on power generation. Through this document, the Brazilian minister of Foreign Affairs, Juracy Magalhães, and the Paraguayan counterpart, Sapena Pastor, agreed to create a Technical Mixed Commission between the two riparians to ensure studies on the feasibility of exploiting the water resources of the basin and preferential energy sales agreements.[13]

Argentina, fearing exclusion from the negotiations and, at the same time, seeing the balance of power definitely hanging in favor of Brazil, proposed a meeting with the five foreign ministers of the riparian countries, which was held in Buenos Aires (Argentina) on 27 January 1967. In fact, Argentina used the opportunity to show its resentment of the Paraguayan-Brazilian project, especially concerning the item IV of the Cataratas Act that prohibited Argentina’s participation in the project, as well as its use of the energy that would be generated at the future plant. On the other hand, the meeting also showed that without the participation of Argentina, any initiative of energy use would be impossible. Argentina also defended its proposal of creating a downstream plant from Itaipu: the hydroelectric of Corpus, in partnership with Paraguay (Pereira, 1974: p. 70).

On 12 July 1977, when the Paraguayan President Alfredo Stroessner was in Rio de Janeiro to conference with his Brazilian counterpart, Gen. Ernesto Geisel (not coincidentally, just ten days before the visit of the Argentine President Jorge Videla to Paraguay) the Brazilian Foreign Minister, Azeredo da Silveira, said that his country could not allow the construction of the hydroelectric plant of Corpus in the quota sought by Argentina, because that would mean giving up national sovereignty.

According to the periodic *Ultima Clave*, seen as the vehicle for the liberal sectors of the Argentine Army, the disagreements over the Parana River would be based on the “hegemonic ambitions” of Brazil in the Southern Cone of America and in its “expansionist strategy based on geopolitical concepts of moving boundaries and manifest destiny”. According to the newspaper: “Itaipu, above all, is a launch platform for the Brazilian hegemony in the Prata Basin”. The criticism from *Ultima Clave* also contemplated the Argentine diplomacy that was not able to define a coherent line of action in face of Brazil’s (quoted in Veja, 1977).

Considering then, that the objectives of Argentina, on one side, and the Paraguayan-Brazilian consortium, on the other, were very different, it soon became apparent that the adoption of common standards for the use of the basin resources would not be easy (Caubet, 2006: p. 124). However, despite the tension, the meeting generated some hopes that outlined the initial construction of a hydropolitical regime in the subcontinent.

Despite the latent scenarios of tension, at the Meeting of Chancellors held in Buenos Aires, a preliminary agenda was set up in which were defined the basic objectives of a joint action for the Prata countries for promoting the regional development. To make it viable, it was decided upon to create an Intergovernmental Coordinating Committee (ICC) in charge of centralizing information on the Basin and coordinating the action of the five countries in partnership with the national water agencies of each (Caubet, 1989: pp. 53–4).

Soon after, in 1968, at the Meeting of the Foreign Ministers held in Santa Cruz de la Sierra (Bolivia), the Santa Cruz de la Sierra Act was signed, establishing the first contours of the legal system of the Prata Basin. During this meeting, approval was given for the statute of the ICC, which was in charge with the responsibility to prepare and present at the next meeting in the following year, the first draft of a treaty with the goal of institutionalizing the hydropolitical relations in the Prata Basin.

Despite the efforts to create mechanisms that would enable the joint management of resources of the Basin, several mishaps arose. Insofar as the negotiations proceeded, it reinforced an anti-Brazilian sentiment in some important Paraguayan sectors, especially amongst academics who claimed the ownership of the Brazilian Island of Itaipu, as seen, set by the document of Cotegipe (Pecequilo and Hage, 2007).

As mentioned, this was an urgent document which lacked a certain geographical precision. However, studies have found that the area was owned by the Paraguayans since before the war of 1865. At that time not only in Paraguay, but in most of South America, Brazil was considered, despite of its status as a great nation, a country indifferent to the fate of its neighbors, without the principle of unity (Pecequilo and Hage, 2007).

Differences and suspicions aside, finally, on 23 April 1969, during the third Meeting of Foreign Ministers, held in Brasilia (Brazil), the Prata Basin Treaty (or Brasilia Treaty) was signed and came into force on 14 August 1970. With broad objectives, the Prata Basin Treaty was a framework agreement that became operational through a series of other agreements and supplementary resolutions resulting from the Meetings of Foreign Ministers that were, in fact, an important decision-making structure within the system of the Prata Basin (Villela, 1984: p. 154).

The first article of the Treaty effectively summarized its general goal. It testified that water can and should be used for various purposes, but no use could be at the expense of other countries, whether of quantity and quality. Thus, we see in the Prata Basin Treaty the birth of an international regime built on sets of principles, rules and decision-making procedures, expressed or implied, around which converges the coordinated agency of the actors that integrate the Basin, towards mutual benefit (albeit still limited and subject to conflicting perceptions about winners and losers). It also showed how politicized water-related issues are in the region and, therefore, the central role that the topic occupies in the regional security agenda.

However, despite the Treaty, many conflicts of interest followed it in relation to the construction of hydroelectric plants, a subject that remained as an extremely touchy point in the international relations of the Southern Cone and, therefore, for the regional balance of power. As Pinto said (2008: p. 45) it was not just energy production, but a latent and continuous confrontation between Brazil and Argentina, with Paraguay playing an extremely important role in this context with its prime geopolitical condition, despite being the economically weakest country in the region.

The Prata Basin Hydroenergetic Structure

The construction of the Itaipu Binational Plant, the result of intensive negotiations that were dragged on for the second half of the 1960s between Brazilians and Paraguayans, made possible the idea of transforming the border dispute between the two countries literally into water. Despite the aforementioned forms of opposition to the enterprise, it can be said that Itaipu found shelter in various sectors in Paraguay, such as civil construction which saw in the project a unique opportunity to leverage the sector.

Overall, the main journals greeted with enthusiasm what was reported as a historical fact. The signing of the Treaty of Itaipu, in 1973, solved two key issues. First, it reassured Paraguay in its challenge regarding the border dispute with Brazil. As the return of the Itaipu Island to Paraguay was out of question, the foreign ministries of both countries decided to give life to a binational enterprise that met the Paraguayan territorial demands as a form of compensation. Thus, on 14 May 1974, it was created the Binational Itaipu, the entity responsible for managing the construction of the plant in the area of dispute. In addition to these diplomatic aspects, Itaipu served economic and geopolitical purposes towards Brazil, giving energy comfort at a time when the country was showing signs of strong growth due to the entry of international capital, and the expansion of its industrial park and infrastructure base (Pecequilo and Hage, 2007).

However, for Argentina the benefits obtained by Brazil would not be only this: as previously mentioned, Buenos Aires perceived with much caution Brazilian politics towards Paraguay, as it strengthened its geopolitical presence in the Southern Cone. Besides that, the Brazilian efforts to make the smaller riparians into satellites in its orbit of influence were, therefore, perceived by Argentina as a decrease in its own power and leverage capacity.

In their efforts to change the situation, Argentina worked with the objective to make an amendment in the Treaty to include the right to prior consultation, leaving to interested parties, including Argentina, to judge the nature and effects of the activities undertaken. Therefore, if there was an indication that the Brazilian-Paraguayan enterprise could cause harm to Argentina, a downstream state, the other riparian should consult it in advance before taking any action. In this sense, an aspect that troubled the Argentines was the Annex B of the Treaty that outlines the general description of the facilities for the production of electricity and adjacent constructions.

After all, the definition of quotas for the energy production from Itaipu could jeopardize its future enterprises, such as Corpus and Apipé-Yacyretá hydroelectric projects, since Argentina would not have any control over the flow of the Paraná River like the upstream countries, especially Brazil. In turn, Paraguay consolidated itself as a true *hydropower emirate* and a strategic partner for both because, in all three cases, it would enter as a partner on equal terms.

In counter-offensive, Brazil rejected the Argentine proposal by suggesting that no country was obliged to supply information in such a way, since it would undermine its national security, its economic development and its efforts to improve the environment. On account of the events described, throughout the 1970s, Brazilian diplomacy has had to manage a major crisis with Argentina, who mobilized itself in various levels, including the General Assembly of the United Nations, to prevent the construction of Itaipu.

Despite all of this, when asked about the issue, the Brazilian chancellor Mario Gibson Barboza said, in 1973, in an interview for *Veja Magazine*, one of the most important vehicles of communication in Brazil, that "more than once I said that a clear, frank and fair friendship between Brazil and Argentina is a historical imperative. I do not think there is any problem between the two countries" (quoted in *Veja Magazine*, 11 April 1973). At least this was the official discourse.

However, in practice, the situation was growing dramatically. The culmination of this tense relationship happened on 31 July 1977, when Argentina, due to the differences between the two countries concerning the exploitation of the Paraná River, had closed its border with Brazil. As we shall see, this deep crisis had a satisfactory ending only in 1979 with the signing of the Multilateral Agreement Corpus-Itaipu, a tripartite treaty that finally normalized the use of Parana River between Argentina, Brazil and Paraguay.

The Tripartite Agreement Corpus-Itaipu

After taking office on 15 March 1979, the Brazilian President, João Batista Figueiredo (1979–1985), announced as one of the priorities of his foreign policy strengthening ties with other Latin American countries, especially with Argentina. For that purpose, Ramiro Saraiva Guerreiro, Secretary-General of the Brazilian Ministry of Foreign Affairs under the previous administration, was appointed as the new chancellor and designated to conduct, with utmost brevity, the resumption of negotiations with Buenos Aires over the exploration of hydropower resources in the Prata basin, seen as essential in order to remove the remaining Argentinean suspicions over Brazil and to prepare, once old tensions had been surpassed, a new scenario based on a strategic partnership that would become possible by the construction of a positive-sum game.

And, indeed, less than seven months after the Figueiredo Administration inauguration, the signing of the "Agreement on Technical Operations between Itaipu and Corpus" by the chancellors Ramiro Saraiva Guerreiro, Carlos Washington Pastor (Argentina) and Alberto Nogues (Paraguay) took place on 19 October 1979, at the Paraguayan city of Ciudad Presidente Stroessner. As stated by the Brazilian Foreign Minister:

> Just in a few months we had signed with Argentina and Paraguay, in October 1979, the Tripartite Agreement on Itaipu and Corpus finally putting an end to this dispute which soured our relations with Argentina during the last eleven years. Without the elimination of this controversy, it wouldn't have been possible to develop the relations with Buenos Aires in the intensity and degree of mutual trust that has characterized Figueiredo's Administration and the successive governments thereafter (Guerreiro, 1992: pp. 91–2).

In other words, the agreement strengthened Brazil-Argentina ties, concurrently, contributing towards a new era, leaving behind a dysfunctional relationship that had dominated the Prata Basin hydropolitical scene, and finally leading two former rivals to revise the rules of the game in their international relations. Thus, with the "latinization" of its foreign policy, successfully fostered by Figueiredo and Saraiva Guerreiro, it was not in the interest of Brazil keeping Argentina a de-industrialized or economically weakened country since, in this context of rapprochement, Brazil needed a strong, dependable strategic partner in future integration projects (Candeas, 2005: p. 33; Oliveira, 2005: p. 199).

Argentina, in turn, had to acknowledge its limitations and the importance of cooperation with Brazil. Supplanted the remnants of rivalry and overcome the tensions caused by the sub-regional (un)balance of power, mutual trust was strengthened; the hypothesis of an armed conflict has become anachronistic and, concomitantly, a new phase was opened in the relations between these two countries, this time marked by the construction of a structurally stable relationship based on cooperation (Candeas, 2005: p. 23).

Paraguay, the greatest beneficiary of the outcome of the situation, ably exploited the complex duality of its strategic pendulous policy confirming, thus, the privileged status of the country that had – and still has – on water resources a powerful political currency.

Therefore, based on the facts mentioned, we may conclude that the Tripartite Agreement Corpus-Itaipu represented a point of inflection between a past of geopolitical disputes and a present of constructive cooperative management of the Prata Basin, thus opening space for a broader and gradual integration project that took place along the following two decades in the Southern Cone of South America. In short, considering the historical events between 1962 and 1991, we have the following timeline for the hydraulic projects developed at the Prata Basin:

Table 6.2 Most Important Hydraulic Projects at the Prata Basin

April 1962	Negotiations between Paraguay and Brazil over the development of the rapids on the River Parana for hydro-electric are interrupted by Brazil, who shows military force, invades and claims control over the Guairá Falls sight.
1967	Brazilian forces withdraw and a joint Brazilian-Paraguayan commission is formed to examine the development of the region.
April 1969	The Prata Basin Treaty is signed by all five riparian states (Argentina, Bolivia, Brazil, Paraguay and Uruguay). The treaty provides a framework for the joint development of the basin and for the management of non-water resources (soil, forest, flora and fauna). An Intergovernmental Coordinating Committee is formed.
April 1973	Itaipu treaty: Brazil and Paraguay announce plans to construct the Itaipu dam; Argentina expresses deep concern for the environmental repercussions of the dam and the its effects on the country's own planned dam project (the Corpus dam).
December 1973	Yacyretá Treaty: a binational organization, Yacyretá, is formed by Argentina and Paraguay to oversee the construction of the hydroelectric dam and the contributing turbines.
1975	Itaipu dam construction begins.
December 1980	Joint declaration is made by the five riparian states expressing a need to promote swift development of the resources on the Prata basin.
September 1989	The riparian states plan to develop the navigational infrastructure of the Parana and Paraguay rivers to make an international waterway navigable. Intergovernmental Commission on the Paraná-Paraguay Hydrovia is formed.
1991	Itaipu hydroelectric plant, the world's largest in power generation (90 billion of kWh), is in full operation after 16 years of construction.

Source: Available at: http://www.transboundarywaters.orst.edu/research/case_studies/La_Plata_New.htm.

The Paraguay-Parana Waterway: Towards Cooperation

While the 1960s and 1970s were marked by a prevailing climate of tension raising doubts about the feasibility of an integration process in the Southern Cone, during the following decades – 1980s and 1990s – an emerging structural stability was established. In the strategic-military field two events confirmed the change of perception in the bilateral relations between Brazil and Argentina: the conclusion of the Agreement on Cooperation for Development and Application of Peaceful Uses of Nuclear Energy (1980) and the Brazilian declaration of support to Buenos Aires during the Falklands War (1982). Subsequently, the spillover effect of this process successfully spread to the economic field reaching its peak in 1991 with the creation of the Southern Common Market (Mercosul).

However, with regard specifically to the hydropolitical contributions, the main multilateral project following the Tripartite Agreement, which signaled an effective physical integration of the Southern Cone, was the Paraguay-Parana

Figure 6.7 Paraguay-Parana Waterway

Source: Copyright © CIC 2011. All rights reserved. Available at: http://cicplata.org/?id=hpp_sintesis.

Waterway (PPW). It represents the most practical and cheapest option for breaking the isolation of the vast South American agricultural frontier formed by the central-western portion of Brazil, eastern Bolivia and northern Paraguay, despite concerns about the likely environmental impacts that might be generated by this enterprise.

As shown in the following map, with a total length of 3,442 km, this waterway links the five basin countries (Argentina, Bolivia, Brazil, Paraguay and Uruguay), beginning in the port of Caceres in Mato Grosso, Brazil, and ending in Nueva Palmira, Uruguay. Its area of influence is 1.75 million km^2 with a population exceeding 17 million: (see Figure 6.7).

Therefore, as an important part of a physical integration project, the PPW would allow considerable reduction in the costs of grain transportation (especially soybeans) by improving intra-zonal trade and the comparative advantages of the South American economic space. Besides, for two countries, Paraguay and Bolivia, the waterway represents a strategic goal: a concrete opportunity to break off the secular mediterranean condition that deprives them of an access to the Atlantic Ocean. For these reasons, according to Sanguinetti (1991: p. 43), the strategic dimension of this project is comparable in terms of importance to the European Coal and Steel Community representing one step further towards the dissipation of possible sources of tensions derived from hydropolitical issues.

Notwithstanding its importance, the project only advanced at the end of the 1980s with the creation of the Intergovernmental Committee of the Waterway (1989) and the following celebration of the Agreement of Las Leñas (1992), the legal instrument that established six basic principles for the sustainable use of the waterway:

a. free traffic;
b. free participation of flags in the trade between member countries;
c. equal treatment to all shipping companies;
d. simplification of customs procedures;
e. safety of navigation; and
f. protection of the environment (Araújo, 2010: p. 39).

It is also noteworthy that, shortly beforehand the Agreement of Las Leñas, in 1991, Resolution No. 238 of the nineteenth Meeting of Foreign Ministers had already included the Paraguay-Parana Waterway project as part of the institutional Prata System (Intergovernmental Coordinating Committee of the Countries of the Prata Basin). Since then, a dozen studies about its economic, financial, environmental, social, legal and institutional developments were completed (Zugaib, 2006).

Despite all the remained uncertainties, particularly environmental ones, in accordance with the principles of Mercosul the waterway is considered an important tool in the strategy of constructing a unified space. The idea that the economic basis of Mercosul presupposes the strengthening of its physical integration has dominated the political discourse and consequently the PPW has been presented as a priority for revitalizing this free trade area. Therefore, the PPW project allows us to affirm that hydropolitics keeps playing a major role in the process of building structural stability in the international relations of South America.

It also strengthens the idea that the Southern Cone moves toward establishing itself as a Pluralistic Security Community, marked by the desecuritization of its thematic agenda and where actors no longer perceive each other as a threat (Deutsch et al. 1957: pp. 1–4). This way, sustained by an extraordinary qualitative improvement in their friendship relations, basin states agreed to identify and prioritize cooperative projects, and to provide the technical and legal structure to see them implemented.

Conclusions

The arguments discussed along these pages demonstrated that international hydropolitics is strongly associated with relations of interdependence established among those actors who share the resources of a water system and/or suffer from transboundary effects of its use, highlighting an intrinsic connection with the logic of adjacency (territoriality). It was verified that hydropolitics have connections with potentially meaningful effects in sensitive issues such as the regional balance of power, the degree of economic development of the riparian states and, therefore, the security of each one of those who depended directly or indirectly, to a greater or lesser extent, on these resources.

The South American Regional Security Complex confirms this assertion by standing out internationally as a great communal water structure because of bearing in its domains important watersheds, among them, the Prata Basin, the world's fourth largest in extension. Therefore, as presupposed by the historical facts, it is natural that it takes a leading role as a strategic issue in the international relations of this region and, concomitantly, in the definition of its security agenda.

Events of the period analyzed in this chapter (1960–1992) demonstrate the validity of the statements mentioned above, the complexity of the subject and also its ambivalent nature. If the years between 1960 and 1979 were marked by a context of conflict and distrust, the following years announced an alternative way for cooperation inaugurating a new stage in the international relations of South America with:

1. the end of the standoff over the construction of Itaipu;
2. the process of redemocratization in Brazil and Argentina;
3. the physical and economical integration, whose main landmark in the field of hydropolitics was the Paraguay-Paraná Waterway project.

So, considering the analysis undertaken, we can conclude that there was a slow but progressive, and welcome, desecuritization of a particularly sensitive issue in the international relations of the Southern Cone. We have seen in this socially constructed anarchy, a prevailing Kantian logic, in other words, the dominant behavior being shaped by cooperation, and the conflicts that arose being solved in a peaceful way. Therefore, the recognition of these practices resulted in a positive-sum game, in which the benefits of cooperation have directly influenced the way that riparians perceived themselves.

Because the logic of hydropolitics is "largely about subsets of actors that are highly interdependent in the face of one river system" (Buzan et al., 1998: p. 199), the Prata Basin demonstrated the validity of the constructivist argument that the identification of threats and the construction of identities among actors within these communal spaces are the result of social processes based on cognitive interpretations and connections originating, in large measure, from historical relations of amity/enmity developed due to use of shared resources of a hydrographic structure.

After all, as indicated by Buzan et al. (1998: p. 18):

> upstream and downstream powers and other potential beneficiaries from a particular river [...] will see each other as both threats and potential allies, which might play into other rivalries and constellations in the region and thus become tied into a more general regional security complex.

Therefore, water and water-related issues were, and definitely are, factors of high importance in the multisectorial dynamics of the Southern Cone of South America. Thus, the analysis presented allows us to identify the existence of a Hydropolitical Security Complex at the Prata Basin, a structural arrangement that prevailed until 1979. Immediately after the resolution of Corpus-Itaipu issue, a linear and progressive desecuritization of the Southern Cone's hydropolitical agenda took place in line with a series of other emblematic factors that contributed towards the establishment of a strategic partnership based on the institutionalization of a power-sharing project (positive-sum game) that laid the foundations of a new stage in South American international relations.

In conclusion, here is the synthesis of the *rationale* that defines the theoretical model of the Hydropolitical Complex of the Prata Basin, a component of a Regional Security Complex of the heterogeneous type. Marked by symbiotic ties established between the ontological elements that make up such a structure, it is the result of a remarkable degree of hydrological interdependence that inevitably connects them into a multisectorial and highly politicized plot, the result of a prevailing context of cooperation observed mainly in the 1980s when there was a significant removal of the constraints originating from a long period of antagonism and mutual distrust.

Notes

1 The study of hydropolitics is a relatively new academic pursuit and the term is used here to denote the set of actions for the identification and resolution of claims resulting from the perception that water, and water-related issues, act as an important component for the geopolitical stability of a given space (for example, riparian states or regions) (Dinar, 2002: p. 229).

2 Once watersheds are fixed territorial structures, it is easy to verify that states, due to this same reason, stand out as the main actors in water geopolitics. Although the focus of hydropolitics converges towards the classical state-dominated model, this fact does not dismiss the importance of non-state actors as, for example, societal communities whose identity and/or development are directly linked to the shared water resources.

3 In 'Security: A new framework for analysis' (1998), Buzan, Wæver and de Wilde propose the expansion of the security agenda to allow the inclusion of threats other than just those coming from traditional political and military fields, but also those stemming from the environmental, societal and economic sectors, each one with its own dynamics, however, keeping the attributes that qualify a given issue as a matter of security:

   1) threats perceived as a potential risk to the survival of a referent object,
   2) the emergency nature of a situation,
   3) state power claiming the legitimate use of extraordinary means, including military force, to address the problem (Wæver, 1994: p. 3).

4 According to the categorization of speech-acts as proposed by Austin (1990), the elocutionist act would be 'what is said', the use of linguistic signs (sounds and words) with meaning; illocutionary acts refer to 'how something is said', they are specific sentences pronounced in a given circumstance and with a purpose; and finally perlocutionary acts that are the concrete effects of saying something. It also should be noted that this typology is a theoretical abstraction once that the linguistic act must be considered as a whole (Austin, 1990: p. 121).

5 When interactions occur between the various levels of analysis – domestic, regional, interregional and global – reflecting all possible combinations among actors and sectors in the security area, we have Security Constellations. This concept, broader than the Regional Security Complex, allow us to separate these four levels, for purposes of analysis, and then aggregate them again in order to provide an overview of the whole. They do reflect the totality of possible security interrelationships at all levels (Buzan et al. 1998: p. 201).

6 Hence, the conceptual use of region by the group of Copenhagen as a territorially coherent space comprising two or more states, once the regular presence of this unit of analysis in any sector offers consistency and density to the model of the Regional Security Complex, especially in sectors such as the political, military and societal ones where the adjacency is a determining factor in the security dynamics.

7 As defined by Deutsch et al. (1957), Security Communities are composed of states that share fundamental values and behaviors to adapt their principles, rules and common institutions and processes of joint decision on behalf of peaceful coexistence. This concept was recovered by Adler and Barnett (1998) that incorporated constructivist elements and the role of shared identities. These communities may form a new political entity (amalgamated communities), as occurred in the process of German unification, or keep its members as independent units (plural communities), such as the European Union. In both cases, the security is notably marked by processes of desecuritization or, in other words, by a Kantian social structure where the actors stop perceiving themselves as threats.

8 A standard security complex is one "that does not contain a global level power, and whose local polarity is defined wholly by the regional powers within it" (Buzan and Wæver, 2003: p. 492).

9 Some Regional Security Complexes (RSC), usually those in which there are a large number of actors, contain intermediaries arrangements called Subcomplex. These sub regional structures have essentially the same attributes that define a RSC, however, their patterns of security interdependence differ from each other.

10 It is responsible for supplying 20 percent of the total energy consumed in Brazil and 92 percent of the energy demanded in Paraguay.

11 The Paraguayan War had its beginning in 1864, as a result of the ambition of dictator Francisco Solano Lopes who had as a goal to increase the Paraguayan territory and get an exit to the Atlantic Ocean through the rivers of the Prata Basin.

12 The Baron of Cotegipe, after the Paraguayan War, established an emergency protocol as a transitory solution to the border disputes between the warring parties. Therefore, there was a need to establish a subsequent document to definitively solve the question of borders between Brazil and Paraguay. However, despite its importance, during the first part of the twentieth century the problem remained with some degree of uncertainty, thus postponing its definitive resolution.

13 Both documents, The Iguaçu Act and the Itaipu Treaty, ensured that the energy not used by one of the joint owners could only be negotiated with the other, at fair prices. As Paraguay uses only 5 percent of its amount, all the remaining energy is sold to Brazil.

References

Adler, E. 2006. Constructivism and international relations, in *Handbook of International Relations*, edited by W. Carlsnaes, T. Risse and B.A. Simmons. London: SAGE Publications, pp. 95–118.

Adler, E. and Barnett, M. 1998. *Security Communities*. Cambridge: Cambridge University Press.

Allan, J.A. 2001. *The Middle East Water Question: Hydropolitics and the Global Economy*. London: I. B. Tauris & Co.

Allan, J.A. 2002. Water resources in semi-arid regions: Real deficits and economically invisible and politically silent solutions, in *Hydropolitics in the Developing World: A Southern African Perspective*, edited by Anthony, R.T. and Roland, H. Pretoria: African Water Issues Research Unit (AWIRU), pp. 23–36.

Araújo, J.T. 2010. Infraestrutura e integração regional: o papel da IIRSA [Infrastructure and regional integration: The role of IIRSA]. *Política Externa*, 18(3), pp. 33–48.

Austin, J.L. 1990. *Quando dizer é fazer: palavras e ação* [*When Saying is Doing: Words and Actions*]. Porto Alegre: Artes Médicas.

Brazilian National Water Agency. 2001. *Bacias brasileiras do Rio da Prata: Avaliações e propostas* [*Brazilian Basins of the River Plate: Assessments and Proposals*]. Brasília: ANA.

Buzan, B. 1991. *People, States and Fear: An Agenda for International Security in the Post-Cold War Era*. Chapel Hill: University of North Carolina.

Buzan, B. and Wæver, O. 2003. *Regions and Powers: The Structure of International Security*. Cambridge: Cambridge University Press.

Buzan, B., Wæver, O. and de Wilde, J. 1998. *Security: A New Framework for Analysis*. Boulder and London: Lynne Rienner Publishers.

Candeas, A.W. 2005. Relações Brasil-Argentina: uma análise dos avanços e recuos [Brazil-Argentina relations: An analysis of advances and retreats], in *Revista Brasileira de Política Internacional*, 48(1), pp. 178–213. [Online]. Available at: http://www.scielo.br/pdf/rbpi/v48n1/v48n1a07.pdf.

Castro, J.E. 2007. Water governance in the twentieth-first century. *Ambiente e Sociedade*, 10(2), pp. 97–118.

Caubet, C.G. 1989. *As grandes manobras de Itaipu: Energia, diplomacia e direito na Bacia do Prata* [*The Great Maneuvers of Itaipu: Energy, Diplomacy and Law at the Prata Basin*]. São Paulo: Acadêmica.

Caubet, C.G. 2006. *A água doce nas relações internacionais* [*Freshwater on International Relations*]. São Paulo: Manole.

Clarke, R. and King, J. 2005. *O Atlas da Água* [*The Atlas of Water*]. São Paulo: Publifolha.

Cosgrove, W.J. 2003. *Water Security and Peace*. A synthesis of studies prepared under the PCCP–Water for Peace Process (a UNESCO–Green Cross International Initiative). Technical Documents in Hydrology, PC-CP n.29. Paris: UNESCO. [Online]. Available at: http://unesdoc.unesco.org/images/0013/001333/133318e.pdf.

Dinar, S. 2002. Water, security, conflict and cooperation. *SAIS Review*, 22(2), pp. 229–53.

Deutsch, K.W., Burrell, S.A., Kann, R.A. 1957. *Political Community and the North Atlantic Area*. Princeton: Princeton University Press.

Elhance, A.P. 1999. *Hydropolitics in the Third World: Conflict and Cooperation in International Rivers*. Washington: US Institute of Peace.

Fajardo, J.M.C. 2004. *Acordo Tripartite Itaipu – Corpus: Ponto de inflexão entre a disputa geopolítica e a cooperação* [*Tripartite Agreement Itaipu – Corpus: Inflection Point between Geopolitical Struggle and Cooperation*]. Porto Alegre: University of Rio Grande do Sul.

Ferreira, L. 2003. Ano internacional da água doce: Repensar a hidropolítica no contexto da segurança regional (International year of freshwater: Rethinking hydropolitics in the context of regional security). *Negócios Estrangeiros*, 6, pp. 39–60.

Filho, A.G.Y. 2006. *Aproveitamento hídrico na Bacia do Prata: Entre o conflito e a cooperação* [*Hydropolitics at the Prata Basin: Conflict and Cooperation*], third ANPPAS Meeting, Brasília, Brazil, 23 May 2006.

Gleick, P.H. 1993. Water and conflict: Fresh water resources and international security. *International Security*, 18(1), pp. 79–112.

Gleick, P.H. 2000. *Water Conflict Chronology*. Oakland: Pacific Institute for Studies in Development, Environment and Security.

Guerreiro, R.S. 1992. *Lembranças de um empregado do Itamaraty* [*Memories of an Employee of the Itamaraty*]. São Paulo: Siciliano.

Hinrichsen D., Robey, B. and Upadhyaya, U. 1998. Rumo a uma evolução azul – Questões globais: Águas ameaçadas [Towards a blue evolution – Global issues: Threatened waters]. *Revista eletrônica da Agência de divulgação dos Estados Unidos*, 4(1). [Online]. Available at: http://usinfo.state.gov/journals/itgic.

Jervis, R. 1982. Security Regimes. *International Organization*, 36(2), pp. 357–78.

Johson, J.B. and Joslyn, R.A. 1991. *Political Science Research Methods*. Washington: Congressional Quarterly Press.

Keohane, R.O. 2005. *After Hegemony: Cooperation and Discord in the World Economy*. Princeton: Princeton University Press.

Keohane, R.O. and Nye, J. 1989. *Power and Interdependence*. New York: Harper Collins Publishers.

Kistin, E.J. 2007. *Transboundary Cooperation in SADC: From Concept to Implementation*. 8th WaterNet/WARFSA/GWPSA Symposium, Lusaka, Zambia, 30 October to 3 November 2007.

Kratochwil, F. 1989. *Rules, Norms and Decisions*. Cambridge: Cambridge University Press.

Le Prestre, P. 2000. *Ecopolítica Internacional* [*International Ecopolitics*]. São Paulo: Senac.

Libiszewski, S. 1995. *Water Disputes in the Jordan Basin Region and Their Role in the Resolution of the Arab-Israeli Conflict*. Zurich: Center for Security Studies and Conflict Research.

Milliken, J. 2001. Discourse study: Bringing rigor to critical theory, in *Constructing International Relations: The Next Generation*, edited by K. Fierke and E. Jorgensen. New York and London: M. E. Sharpe.

Ohlsson, L. 1995. Introduction: The role of water and the origins of conflict, in *Hydropolitics: Conflicts Over Water as a Development Constraint*, edited by L. Ohlsson. London: Zed Books.

Oliveira, H.A. 2005. *Política Externa Brasileira* [*Brazilian Foreign Policy*]. São Paulo: Saraiva.

Pecequilo, C. and Hague, J.A. 2007. Dilemas sul-americanos: o Brasil, o Paraguai e Itaipu. [South American dilemmas: Brazil, Paraguay and Itaipu]. *Meridiano 47*. [Online]. Available at: http://meridiano47.info/2007/07/19/190720071526/.

Pereira, O.D. 1974. *Itaipu: prós e contras* [*Itaipu: Advantages and Disadvantages*]. Rio de Janeiro: Paz e Terra.

Pinto, T.G. 2008. *Itaipu: Integração em concreto ou uma pedra no caminho?* [*Itaipu: A Concrete Integration or a Stone on the Way?*]. São Paulo: Manole.

Procópio, A. 2007. *Subdesenvolvimento Sustentável* [*Sustainable Under-development*]. Curitiba: Juruá.

Salman, S.M.A. and Boisson de Chazournes, L. 1998. *International Watercourses: Enhancing Cooperation and Managing Conflict*. Washington: World Bank Technical Paper n. 414.

Sanguinetti, J. 1991. Proyecto Hidrovía: Oportunidades del empresariado [Waterway Project: Entrepreneurship Opportunities]. Integración Latinoamericana, *INTAL*, 168, pp. 42–51.

Schilling, P.R. 1981. *O expansionismo brasileiro: A geopolítica do General Golbery e a diplomacia do Itamaraty* [*The Brazilian Expansionism: Geopolitics of General Golbery and the Diplomacy of Itamaraty*]. São Paulo: Global.

Schulz, M. 1995. Turkey, Syria and Iraq: A hydropolitical security complex, in *Hydropolitics: Conflicts Over Water as a Development Constraint*, edited by L. Ohlsson. London: Zed Books.

Searle, J.L. 1984. *Os atos da fala: Um ensaio de filosofia da linguagem* [*Speech-Acts: An Essay*]. Coimbra: Almeida.

Senghaas, D. 1988. *Konfliktformationen im Internationalen System* [*Conflict Formations in the International System*]. Frankfurt: Suhrkamp Verlag.

Sheenan, M. 2005. *International Security: An Analytical Survey*. Boulder and London: Lynne Rienner.

Sironneau, J. 1996. L'eau: Nouvel enjeu stratégique mondial [Water: New global strategic challenge]. *Coll. Poche Geopolitique, Economica*, 6, Paris.

Stahl, B. 2007. *Who Securitized What, When, and How? A Comparative Analysis of Eight European Union Member States in the Iraq Crisis*. Paper presented at the ECPR conference in Turin, September 2007.

Tundisi, J.G. 2003. *Água no século XXI: enfrentando a escassez* [*Water in the Twentieth-first Century: Facing the Shortage*]. São Paulo: RiMa, IIE.

Turton, A.R. 2001. *Hydropolitics and Security Complex Theory: An African Perspective*. 4th Pan-European International Relations Conference, University of Kent, Canterbury, 8–10 September, 2001. Available at: http://www.up.ac.za/academic/libarts/polsci/awiru.

Turton, A.R. 2003. *The Political Aspects of Institutional Developments in the Water Sector: South Africa and its International River Basins*. Pretoria: University of Pretoria. [Online]. Available at: http://upetd.up.ac.za/UPeTD.htm.

Turton, A.R. 2008. A South African perspective on a possible benefit-sharing approach for transboundary waters in the SADC region. *Water Alternatives*, 1(2), pp. 180–200.

United Nations Educational, Scientific and Cultural Organization (UNESCO). 2003. Water for people, water for life. *UN World Water Development Report*.

Van Ginkel, H. 2001. Discourse of the United Nations Under-Secretary General at the Stockholm Water Symposium, 13 August 2001. Quoted in *Financial Times*, 14 August, 2001, 6.

Väyrynen, R. 1984. Regional conflict formations: An intractable problem of international relations. *Journal of Peace Research*, 21(4), pp. 337–59.

Veja. 1973. Um entendimento inevitável [An inevitable understanding]. 11 April 1973. Available at: http://veja.abril.com.br/arquivo_veja/capa_11041973.shtml.

Veja. 1977. A sofrida procura da paz [The painful search for peace]. 27 July 1977. Available at: http://veja.abril.com.br/arquivo_veja/capa_27071977.shtml.

Villela, A.M. 1984. O Tratado da Bacia do Prata [The Prata Basin Treaty]. *Revista de Informação Legislativa*, 21(81), pp. 147–76.

Viola, E. 2005. *A hegemonia das democracias de mercado no sistema internacional do início do século vinte e um* [*Hegemony of the Market Democracies in the Twentieth-first Century*]. Brasília: Universidade de Brasília.

Wæver, O. 1994. Securitization and desecuritization, in *On Security*, edited by R. Lipschultz. New York: Columbia University Press.

Waltz, K. 1959. *Man, the State and War*. New York: Columbia University Press.

Wendt, A. 1999. *Social Theory of International Politics*. Cambridge: Cambridge University Press.

Wolf, A.T. 1998. Conflict and cooperation along international waterways. *Water Policy*, 1(2), pp. 251–65.

Woods, N. 1996. *Explaining International Relations Since 1945*. Oxford: Oxford University Press.

WWF. 2007. *World's Top 10 Rivers at Risk*. Gland: WWF International.

Yetim, M.E. 2003. Governing international rivers of the Middle East. *World Affairs*, Fall, 2003.

Yoffe, S., Fiske, G., Giordano, M., Giordano, M., Larson, K., Stahl, K. and Wolf, A.T. 2004. Geography of international water conflict and cooperation: Data sets and applications. *Water Resources Research*, 40(5), pp. 1–12.

Zugaib, E. 2007. *A Hidrovia Paraguai-Paraná* [*Paraguay-Paraná Waterway*]. Brasília: FUNAG. Available at: www.funag.org.br.

Chapter 7
A Need for New Institutions? Bilateral Deals over Food Producing Resources and International Co-Operation on Food Security

Benjamin Shepherd

Introduction

Following dramatic food price rises during 2007/08, and with the number of people living with chronic hunger exceeding one billion people in 2009 (FAO, 2009c), the importance of food security on the international agenda continues to rise. As a looming crisis of the modern world with significant implications for the well-being of all peoples, food security is gaining the attention of world leaders and global institutions. Many states are starting to look towards securing their future food supplies, and one aspect of this, which has gathered increasing criticism from observers, is the growing phenomenon of "grabs" of agricultural land in the developing world by certain countries. The purpose of these deals is usually to secure food production for export back to these countries' domestic markets. These deals potentially provide an opportunity for investment and employment in the developing countries, but many of the commentators criticise them as a modern form of colonialism.

Until recently, a major concern has been the absence of any meaningful international frameworks for ensuring these bilateral arrangements will deliver desirable outcomes for the developing country parties. At the 2009 World Food Security Summit in Rome (FAO, 2009e), a number of multilateral organisations including the Food and Agriculture Organisation (FAO) of the United Nations (UN), the International Food Policy Research Institute (IFPRI) and the World Bank came together in support of the idea of a *Code of Conduct* (the Code) to address this lacuna.

The purpose of this chapter is to critically analyse these multilateral organisations' response to the investment deals via the Code; in particular how they respond to the criticisms levelled at the deals and the risks the deals create for all stakeholders. To this end, the first section discusses the nature of the deals, the forces behind them and the risks arising from the deals. In the process it identifies a pair of competing narratives that represent alternate perspectives on the deals from the points of view of the investors and their supporters against that of the deals' detractors.

The second section then examines the *Code of Conduct* as the primary response by these international organisations and argues that the Code is weak as a proposed mechanism for addressing the deals for legal and political reasons. But further, and importantly, it also argues that the international organisations effectively use the Code to legitimise one set of interests portrayed in the competing narratives to the denial and exclusion of the other set of interests. This is important because it demonstrates a number of significant failures of the international organisations in responding to the deals. Firstly, it demonstrates partisanship on the part of the international organisations to the interests of national governments and trans-national financial capital. Secondly, it nullifies the valid concerns put forward in the opposing narrative. Finally, it encourages the drivers behind the deals in the first place to be ignored and left unaddressed.

In identifying these failures of the international organisations as illuminated by the Code, this paper alleges that the international organisations are not adequately, competently or fairly addressing either the deals themselves or the forces that have given rise to them. In conclusion it is proposed that consideration of alternative approaches is necessary for maximising protection of the weaker parties in such deals at the same time as enhancing the long term food security of the states and communities involved to avoid the need for such arrangements at all.

Food Security and Land Deals

Historically, a states' desire for land has been pursued by war or colonisation. Over the last half century these strategies have become increasing less tolerated in the international order and today more unobtrusive, economic measures are deployed. That said, exploitation of foreign lands for profit and to export food supplies home is nothing new and does not go uncriticised: the British East India Company and America's United Fruit Company in Latin America might have been a century apart but both have become bywords for historic cases of corporate pillage of foreign lands (see for example Davis, 2001; Chapman, 2008). Today, this new wave of land acquisitions is often justified by pursuit of "food security". Food security is itself a part of the new security frontier, but the pursuit of foreign lands by states seeking food security adds a significant dimension to the contemporary security *problematique.*

The term food security is used widely but has supposedly been given some 200 different definitions (Shaw, 2007: 383) suggesting it means many different things to many different people. In general, most contemporary applications of the term are consistent with the Food and Agriculture Organisation (FAO) of the United Nations (UN), which as the peak international organisation tasked with the responsibility of overseeing global food security, defines it such that:

> Food security exists when all people, at all times, have physical, social and economic access to sufficient, safe and nutritious food that meets their dietary needs and food preferences for an active and healthy life. (FAO, 2009d)

For the FAO, which is the peak international organisation tasked with the responsibility of overseeing global food security, this definition is essentially an aspirational mission statement in the effort to reduce hunger. This definition locates the concept as primarily an economic one: It defines food security as a problem of supply (of sufficient, nutritious food) and demand (to meet the dietary needs of all people at all times) and how supply is made available to meet demand (by physical, social and economic means of access). Indeed, the term food security is most frequently used as an economic concept to describe the challenges of providing food for a society; a challenge based on competition for scarce resources in an increasingly resource constrained world. But food security is also a technical term used by agricultural scientists in terms of developing technologies which enhance crop productivity or increase crop resilience to shocks such as drought, flood and plague[1] and by commercial enterprises in the exploitation of same.[2] It is used in international law particularly in relation to international humanitarian law and human rights, of which the "Right to Food" is one of the most fundamental and least contentious.[3] It is even used by practitioners of the health sciences in relation to nutrition.[4] It is also a political tool used to justify pursuit and control of resources, to legitimise action or policy and to further political agendas.[5]

In addition, food security is a *security* challenge. It is a security challenge for three reasons. First, when failures in improving economic food security—particularly from competition over scarce resources and/or sudden collapse of supply—have the potential to engender conflict, or even violence, as was seen in the reactions to the food price shocks in 2007 and 2008; just the latest chapter in the common story of "food riots" throughout history. Second, it is a security challenge because widespread chronic hunger is a form of structural violence (after Galtung, 1969): that is, the infliction of avoidable physical harm against human beings but by institutional, rather than personal, forces. Third, and most importantly in this context, it is a security problem for a regime that needs to actively work to maintain food security to retain power and then further still when the consequence of those actions have repercussions for other actors. In all three ways, these land deals are an exemplar of new security challenges.

In practice, for many governments food security is nonetheless treated as an economic challenge—seeking scarce resources to guarantee adequate availability of affordable food for their populations. However, the implication for a government that is unable to meet this challenge is the risk of deprivation of at least some of its citizens and the loss of legitimacy and propensity to violence that this deprivation potentially engenders. Protests, civil unrest and conflict erupted in as many as 30 countries from Haiti to India in reaction to dramatic rises in staple foodstuffs during 2007 and 2008. Prices of food staples rose by up to as much as 400 percent for corn and 200 percent for rice by some estimates (e.g. BBC, 2007; Nicoll, 2008; Steinberg, 2008) and the widely reported upheavals provided salutary examples of political violence resulting from sudden deprivation of food staples for many (see, for example, Nicoll, 2008; Brown, 2008a; Guyler and Loney, 2008; Taylor and Buncombe, 2008; BBC, 2007; BBC, 2008a).

Since these events, some governments appear to have become increasingly anxious about being able to meet the growing demands of their populations. This is evidenced by the emergence of new policies to try and secure long-term sources of supply. One feature of these new policies is a move to invest (or actively support investment) in substantial parcels of foreign agricultural lands for the purposes of producing food for shipping back to their domestic markets. This is a policy that is being pursued by some economically important countries including China, South Korea, Saudi Arabia and the other Arabian Gulf states. Although these are policies instituted for domestic purposes, they are of international concern for two important reasons. First, because they by-pass the international market as the source of non-domestically produced foodstuffs. Second because they largely target farmlands in developing countries, often with weak governance and institutional capacity to effectively handle the influx of foreign participation in local agriculture.

Although neither the acquisition of agricultural land in overseas jurisdictions nor Foreign Direct Investment (FDI) in poorer countries are new undertakings, this focus on developing country lands is one of three characteristics that seem to make these particular "food securing" deals remarkable. The first characteristic is that many of the states targeted by investors rely on food aid to feed large numbers of people in parts of the country. This includes Sudan, DR Congo, Ethiopia, Kenya and Uganda which are five of the top ten beneficiaries from the World Food Program (WFP, 2009). Moreover, at least nine out of the 32 countries currently on the FAO's *Countries in Crisis* watch-list (FAO, 2009a) are reported to be either hosts to, or targets of, this type of foreign agricultural investment. Another three known hosts of large foreign land investments appear on FAO's appended list of nine "countries with unfavourable prospects for current crops". Most of these states targeted by investors have little economic power, weak governance and low institutional capacity to manage the foreign investors.

The deals are also notable in terms of scale and proliferation. In 2009, IFPRI estimated 20 million hectares—or as *The Economist* (2009) described it, "the size of France's agricultural land and a fifth of all the farmland of the European Union"—were acquired or under negotiation during 2007 and 2008.

Third, and perhaps most importantly, the deals are unusual in terms of the explicit involvement of the investing country governments in the deals and their use of "food security" as the reasoning and justification for the investments. In many cases, the foreign investment in agricultural land has become official government policy for improving domestic food security: for example China (People's Republic of China, 2008: Section 4(c)); Saudi Arabia (KSA MOFA, 2010); Qatar (for example, as reported by England and Blas, 2009; and Walid, 2009) and Kuwait (Finance Minister al-Shamali cited in Reuters, 2009; following earlier reports such as Reuters, 2008a).

Additionally, the funding for many of these investments is coming directly from the governments. In the case of China for example, much of the activity in foreign jurisdictions appears to be undertaken by para-statal enterprises (Raine, 2009; Horta, 2008). In the case of the Gulf state investors, the governments are either making Sovereign Wealth Fund (SWF) resources available, or have established funds that commercial organisations (both privately owned and publicly listed) can draw funding from. Saudi Arabia, for example, has reportedly established a 20 billion Riyal (US$5.3 billion) Agricultural Development Fund (ADF) (*Saudi Gazette*, 2010) in conjunction with the *King Abdullah Initiative for Agricultural Investment Abroad* which is the Saudi government's policy cornerstone for the pursuit of food security via foreign farmlands.

Drivers Behind the Deals

Countries like Saudi Arabia and China are instigating policies to acquire agricultural land apparently because they are increasingly concerned about the challenge of providing long term food security for their populations. In the cases of China and Saudi Arabia, challenges arise on both the demand and supply sides of the economic equation. On the supply side for China these range from soil and water degradation in key agricultural areas to urban expansion that is rapidly consuming previously-productive agricultural lands (People's Republic of China, 2008: Preface). On the demand side, even though China has low population growth rate, it has a very large population to feed and a rapidly expanding middle class demanding foods which place increasingly greater strain on limited agricultural production (Kirby, 2009; Bhalla, 2007).

In the case of Saudi Arabia the supply and demand side pressures are manifested differently. On the demand side, Saudi is experiencing rapid population growth, growing from 20 million in 1995 to an expected 36 million by 2025 (US Bureau of the Census, 2010); is highly urbanised, 88.5 percent of the population lived in cities in 2005 (UNEP, 2005);[6] and, has a population with reasonably high expectations—from decades as a generous welfare state (for example, Okruhlik, 1999)—that are being challenged by high unemployment[7] and increasing rates of poverty (Shaheen, 2009; al Arabiya, 2009). On the supply side, the Kingdom has invested hugely in the pursuit of domestic agriculture—greening the desert—but with high economic, environmental and social costs which have made alternative sources of supply imperative. Notwithstanding its domestic agriculture, Saudi Arabia is highly dependent on imports for the majority of its food supply (FAO, 2002; Abdullah, 2010).

For Saudi Arabia, the imposition of export bans during the 2007/08 food price shocks by some of its most important trading partners—including India, its primary provider of one of its most important staples, rice (see BBC, 2008b; Reuters, 2008b; *China Post*, 2008; Oryza, 2008; Taylor and Buncombe, 2008)—was a major surprise that appeared to stimulate the dramatic shift in policy towards the acquisition of foreign farmlands. This apparently alarmed Saudi policy makers (RCCI, 2008) and preceded the instigation of the foreign land investments as government policy. It points to a decreasing confidence in the international market as a trustworthy mechanism to meet Saudi's long term food needs.[8]

Competing Narratives

The rapid increase in large scale agricultural land investments in developing countries by China, Saudi Arabia and others has become the subject of vociferous international debate. The debate essentially consists of two competing narratives: On one hand, there has arisen a widely-promulgated, legitimizing narrative around these investments as offering significant developmental outcomes for the host country including improved productivity, the provision of new infrastructure, the securing of land titles and the generation of employment and income. This has basically arisen in reaction to the opposing narrative which describes the deals as "a new form of colonialism" with the likely outcomes of further exploitation of the already poor and disempowered communities whose land is being "grabbed" by the foreign investors for their own gain.

The legitimising narrative articulates the value proposition for both the investor and the host as a "win-win" scenario and is shared not only by the investing countries but also by the host country elites including Tarek Shalabi, Sudan's Minister of Finance (Kielburger, 2009) and Minister of Agriculture, Dr Abdul Halim Ismail Al-Mutaafie (*Sudan Tribune*, 2010); Republic of Congo President Denis Sassou-Nguesso (Parsons, 2009); Zambian Foreign Minister Kabinga Pande (UK Zambian, 2010); and Filipino Secretary for Agriculture Arthur Yap (RP, 2009). All of whom are vocal in encouraging the investments and actively facilitating the availability of land in their countries for the foreigners.

This "win-win" narrative has also been adopted by investment corporations from the global North, who appear to have "jumped on the bandwagon" of the financial opportunities offered by developing country farmlands with low levels of industrialisation. Agriculture has become an increasingly competitive field for wealthy western investors and most appear to be propagating the "win-win" narrative as justification. Examples include: New York's Jarch Capital who in 2009 reportedly acquired 800,000 hectares in deals with warlords in Southern Sudan (Lavallee, 2009; *Sudan Tribune*, 2009) which is over 4 percent of Sudan's total arable land;[9] Australia's BKK, led by former Australian Government Treasurer Peter Costello, is reported to be operating an investment program of some US$600 million in Cambodian farmlands (Socheat, 2010; Finch, 2010) which is FDI to the equivalent of Cambodia's total annual Overseas Development Aid (ODA) (Chanboreth and Hach, 2008); and London fund Emergent Asset Management (EAM), who claim to operate the largest investment fund in African agriculture, which offers returns of:

> +25% per annum from commodity production yields and land price appreciation … yield enhancement … based on the introduction of modern farming techniques and technologies … while agglomerating farms to increase efficiency and generate economies of scale. (Emerging Asset Management, 2010)

Framed by EAM as "win-win" by leveraging:

> Sustainable agricultural practices reflect(ing) the (fund) manager's sensitivity to the environment … (and) includes provisions for carbon credit generation and other "greening" initiatives. Social responsibility is also a key tenet of the strategy and brings economic uplift to (local) communities … (Emerging Asset Management, 2010)

However, EAM are far from forthcoming on *how* the agricultural investments will deliver both the stunning >25 percent per annum return to wealthy (the minimum investment is £500,000) investors at the same time as delivering sustainable agricultural solutions and socially responsible development of poor, local communities. There are scant details on "how" these will be achieved; no guarantees made to meeting these promises and little likelihood that the delivery of promised benefits to the local communities will take priority over maximising investor returns.

Thus it is not surprising that, in the alternative discourse, the deals have come under particular criticism for being a "modern form of colonialism" (for example see Northern commentators such as Petras, 2008, and Monbiot, 2008, and African observers such as Madoffe, 2009). The deals were largely brought to international attention by Spanish civil society organisation *Grain* in 2008 (Grain, 2008b; Grain, 2008c) who have developed a continually updated repository of the staggering volume of media reports on the deals (Grain, 2009a). *Grain* are primarily an activist organisation for small-holding farmers, and their own representations of the deals are firmly rooted in support of at-risk, rural poor and advocates strongly against the potential for exploitation that the deals present local communities in developing countries.

This advocacy exemplifies the alternative narrative which argues that the foreign investors are arriving to produce food-stuffs (and even bio-fuel inputs) intended largely for export back to the investors' domestic markets and to generate profits for the investors' own pockets while offering few, if any, guarantees for the developing country host. This advocacy is not limited to small northern NGOs such as *Grain* (based in Barcelona) and similar European organisations such as *WeltHungerHilfe* (WeltHungerHilfe, 2010) and *Weltwirtschaft und Entwicklung* (Hoering, 2009), but also high profile international civil society organisations such as the Food First Information Network (FIAN, 2010) and Southern NGOs of international standing such as the Civil Society Agriculture Network (CISANET) (Mhone, cited in IRIN, 2009) and a myriad of local organisations ranging from Licadho of Cambodia (Licadho, 2009), to Kenya Nature and the Sungwaya Welfare Association, both of Kenya (also cited in IRIN, 2009), to the Partners in Melanesia of PNG (cited in Grain, 2010).

This anti-exploitation narrative is in the tradition of the discourse-shaping treatise on neo-colonialism by Ghana's first President Kwame Nkumrah (1965) that defined neo-colonialism in terms of control of weak states by external "economic or monetary means" and of "foreign capital (being) used for the exploitation rather than development of the less developed parts of the world". It focuses particularly on the extractive desires of the neo-colonists for the resource wealth of the colonised. Nkumrah believed that:

> Neo-colonialism is also the worst form of imperialism. For those who practice it, it means power without responsibility and for those who suffer from it, it means exploitation without redress.

Proponents of this narrative advocate for the protection of the communities in the developing countries and spell out risks for those communities arising from neo-colonial exploitation.

These identified risks include: loss of land; loss of employment; loss of income and loss of food security; further disempowerment; enslavement to foreign financial interests and destruction of local environments.

A Broader Assessment of Risks and Threats from the Land Deals

While the anti-exploitation narrative clearly articulates the potential risks for local communities, risks from the land deals actually accrue to the proponents of both narratives. A significant weakness in a country's strategy of "going abroad" to bolster its domestic food security is that the foreign investments in farmlands might well *not* achieve the intended securing of the means of production for the long term. An historic example of this is oil. Countries like Saudi Arabia, Kuwait, Iran and Venezuela have all re-taken ownership of their oil producing resources and/or their extractive monopolies from foreign hands. It is hard to foresee that successful long-term developmental investments in other resources would result in a different outcome. Possibly even riskier for the foreign investor is the situation where ownership and control of the resources is contested by local and foreign interests leading to the same kinds of endemic conflict and violence seen over resources in countries like Nigeria and the massive legal and financial wrangling in Russia.

Even more problematic than the notion of embarking on long term food security strategies destined to failure is the problem for the investor to become a catalyst for, or embroiled in, conflict, violence and political turmoil in the host country. The land deal situations are likely to create situations that are compatible with widely held theories of collective violence including deprivation conflicts and violent challenges to political authority (for example, Gurr, 1970; Nkemdirim, 1977; Homer-Dixon, 1991; Homer-Dixon, 1994).

Foreigners arriving to take control of farmlands from local communities present the possibility of creating deprivation scenarios in that community. For example, the removal of land from its customary users; the failure to deliver on promises of employment and income provision; the construction of infra-structure that benefits a distant elite at the expense of the local communities; or, Blas' (2008) "nightmare scenario of food being exported from fortified farms while the hungry outside clamour to be fed". The fact that the food price shocks in 2007/2008 caused protest, riots and violence around the world reiterates how sudden scarcity can be a source of deprivation resulting in political conflict.

The propensity to political violence, and the problem that deprivation presents, is exacerbated when local governance is weak or has low levels of legitimacy. Questions of legitimacy are raised when a government fails to protect the interests of its citizens. The sale of farmland to foreign interests that removes land, food supplies, income and employment from the local population without providing alternatives is a clear example of a government's failure to protect the interests of its citizens. Weakening legitimacy of government can contribute to political instability that has both internal and external consequences. Internally, weakening legitimacy creates space for political alternatives which can lead to expression of political violence.

Such a challenge to government legitimacy related to foreign land investments has a recent precedent. In Madagascar in 2008/2009, the combination of skyrocketing food prices and the conclusion of a vast 1.3 million hectare deal with the Korean *Chaebol* Daewoo caused protests, demonstrations and street violence against wealthy farmer/land owner President Marc Ravalomanana which were "the last straw" (BBC, 2009a) in a political crisis that led to a military coup that toppled the Ravalomanana government in March 2009. Andry Rajoelina, the self-proclaimed president replacing Ravalomanana, publicly reneged on the Daewoo deal (BBC, 2009b).

Externally, weakening legitimacy can also create problems for the investor country—by the legitimacy of its presence being threatened or by it, or its citizens, being directly targeted—as well as by posing risks to regional stability. There has already been reports of rising ethnic tensions and civil unrest in the Zambesi valley against the influx of Chinese workers (Horta, 2008; Raine, 2009: 138–41). At the same time, in both the cases of Saudi Arabia and China as investors in these deals, it has been argued that the legitimacy of their governments rests in part on being able to ensure their populations continued prosperity (for example, see Weatherly, 2006; or Niblock, 2006). Food security is essential to achieving this, and more than simply protection against hunger, but ensuring the availability of foodstuffs demanded by increasingly wealthy consumers.

There is another confounding factor that these deals in low-capacity countries present. That is the challenge to the sovereignty of the developing host countries from the investments. While the sale of land to a foreign investor does not infringe the sovereignty of the host state in the same confrontational and immediately damaging way that appropriation of the land by force would, there are nonetheless subtle erosions of sovereignty that can take place, especially in a state with limited ability to manage the foreign investment. One type of erosion of sovereignty can occur over territory where the land and the benefits of its productivity are excluded from the host country, which is one of the core concerns of the anti-exploitation narrative around the deals. Another can occur over population, where the investing country relocates sizeable populations of its own people in the host country to exploit the acquired land. Local concerns over the aforementioned influx of Chinese foreign workers in Africa warrant further investigation in this regard. Yet another can occur over political and legal control when the investing country flouts the laws of the host country in its development of the acquired land.

China has been accused of this in Africa in relation to the production of genetically modified (GM) crops. A number of African countries that are targets of agricultural land deals have clear policies against GM agriculture and produce. These include Angola, Zambia and Zimbabwe, all of whom are host of current, or proposed, Chinese agricultural investments. At the same time, China has made clear that its investment in African agriculture will incorporate bio-engineered grains (see, for example, *China Daily*, 2008; Grain, 2008a; Horta, 2008; Grain, 2009b) and is already doing so in at least Tanzania and Mozambique. This is presumably as a primary means of increasing agricultural productivity on the African farmlands. This is despite in 2003 the Tanzanian parliament reportedly blocking the import of GM seeds (Baille, 2003). Such cases beg the question how will African countries reconcile their anti-GM policies with the welcoming of Chinese GM-based agriculture on their land and what diminution of sovereignty occurs if the foreign investors does not need to abide by the host countries' laws; pointing to interesting research questions for political scientists and scholars of international law.

To synthesise the analysis so far, the different, and conflicting, interests identified between the host and investors, combined with what Speildoch and Murphy (2009) call the "lopsided power relations" between them, brings into focus the challenges presented by the deals. The conflict of interests are, on one hand, the powerful investor states are pursuing policies with the intent to *protect their populations from the risk of hunger without which their own stability would be at risk*. This set of interests is represented by the legitimizing narrative which describes the deals as a "win-win" scenario for both the hosts and investors. On the other hand, these deals present challenges for the stability of the weaker, developing country hosts: potentially infringing on their sovereignty, diminishing the legitimacy of their regimes and creating or exacerbating deprivation within their populations which has the potential to increase the propensity to collective violence, whether directed internally or against the foreign investor. This set of interests is represented by the narrative of the deals as a form of neo-colonial exploitation. The unequal power balance in favour of the investors exacerbates the concerns around the capacity of the target countries to adequately manage the foreign investments.

International Organisations' Response

That the sets of interests have coalesced around a pair of competing narratives beg the questions on: how existing multilateral organisations are acting to reconcile these conflicting interests; how they are addressing the forces behind the deals in the first place; and, how they are helping to mitigate the risks that the deals present as a resource scarcity issue, as a potential source of conflict and violence and as a challenge to sovereignty and government legitimacy.

So far, the principal "solution" proposed by the multilateral organisations tasked with addressing global food security—primarily the FAO, but also including International Food Policy Research Institute (IFPRI), the International Fund for Agricultural Development (IFAD) and the World Bank—is a *Code of Conduct*. The Code is intended to provide 'an important mechanism for ensuring that these projects are economically, socially and ecologically sustainable' (Meinzen-Dick and Markelova, 2009: 79). Via the Code, the cooperating institutions hope to mitigate the risks arising from the deals at the same time as facilitating the so-called "win-win" scenario of mutually beneficial outcomes for both the investor and the host. IFPRI, as one of the Code's primary advocates, argues that its force will come from its ability to "prepare local people, host governments and investors for constructive negotiations" (Meinzen-Dick and Markelova, 2009: 79).

Although a final documented version of the Code is yet to be released, it's proposed content has been laid out by the institutions in question in the form of seven *Principles for Responsible Agricultural Investment* (World Bank et al. 2010).[10] These principles are:

- "respecting land and resource rights" (page 2);
- "ensuring food security" (page 6);
- "ensuring transparency, good governance and a proper enabling environment" (page 8);
- "consultation and participation" (page 10);
- "responsible agro-enterprise investing" (page 13);
- "social sustainability" (page 16); and,
- "environmental sustainability" (page 18).

Thus it can be seen that the proposed content of the Code has been articulated in such a way that it responds directly to the criticisms posed in the narrative of exploitation, that is, the risks of:

- loss of rights;
- loss of land;
- loss of food;
- loss of income;
- disempowerment; and,
- the exploitation of local communities and the environment.

Therefore, in theory at least, the Code goes some way to addressing the risks the deals present to local communities in the developing host countries. In articulating these principles as desirable outcomes and as mitigations of the risks presented by the deals, the strength of the Code appears to be its creation of a viable norm in relation to these desirable outcomes around which both sets of interests can congregate.

Legal and Political Limitations

However, notwithstanding the laudable desire to encourage an ideal, normative shift for parties to the deals, a significant limitation of the Code is that is it an inherently feeble legal instrument with little power to reconcile the competing interests and conflicting agendas of the participants. Despite being described as "a mechanism" there are few actual mechanisms possible within it: By definition, a code of conduct is free from regulatory enforcement. It is reliant upon voluntary compliance and self-assessment of conformity and the current early proposals appear to offer scant motivation for adherence. At best, codes of conduct are a framework for promoting desired behaviours by participants but nonetheless from the perspective of international law, they are non-binding (Zamora, 1999: 260).

That a non-binding instrument has been the primary response of the multilateral organisations involved speaks to the constraints on international bodies in responding to problems of this type and developing effective legal instruments in such circumstances. These constraints arise in part from the nature of multilateral organisations as largely consensus-based institutions meaning that negotiating enforceable instruments that are interpreted by participants of favouring others over themselves is notoriously problematic. The Kyoto Protocol and Copenhagen negotiations are possibly the most famous contemporary example of this conundrum (for example, see discussion by Vezirgiannidou, 2010).

Additionally, multilateral organisations tend to be constrained to dealing with matters pushed by their powerbrokers or agreed by their members as warranting attention. In the case of the land deals this is particularly difficult because the deals are bilateral in nature and, when agreed as a treaty between two governments or a government and a foreign corporation, fall well outside the scope of warranting multilateral engagement.[11]

That the primary force of the Code is expected to arise from effectively preparing the parties for "constructive negotiations" also has its problems. This approach to using the Code in practice does little to overcome its lack of enforceability. More importantly, it elides the power imbalances between the actors. Even assuming that local communities have access to the Code and gain access to the negotiations, it is unable to ensure those interests will be treated equally by the other parties whether governments or profit-motivated corporations.

This points to two problems for the international organisations, first in influencing non-member entities such as trans-national corporations and second in balancing the interests of their members (i.e. states) with the objective of protecting the disempowered. The FAO has no remit with regard to trans-national corporate interests which are regulated solely by the environments in which they operate (Sornarajah, 2004). If the sponsoring government either owns, or declines to regulate, its corporations, there is little the FAO and its sister institutions can do. If the host countries have inadequate regulation or capacity to enforce regulations pertaining to the investing companies, again the international organisations have no ability to act in response.

Moreover, international organisations such as the FAO, CGIAR and IFPRI are funded by contributions from member states and/or other international organisations, such as the World Bank, of which states are also paying members. Land deal investor countries China and South Korea are both in the top ten contributors to the FAO and Saudi Arabia is in the top 25 (FAO, 2010b). Even host country governments like Angola and Zambia pay token dues—although often to the tune of one thousand times less than the payments from the major contributors[12]—while, of course, the local communities in the host countries do not. This puts the international organisations into a position of weakness when attempting to redress imbalances of power between these parties.

This conflict of interest is not just a theoretical problem for the World Bank. On one hand, the World Bank has been a leading contributor to the development of the seven principles behind the Code. On the other, the World Banks' International Finance Corporation has been reportedly actively involved in funding deals (Mittal, 2010).

The Code and the Competing Narratives

The World Bank's apparent duplicity in turn points to what is perhaps the biggest problem for the Code: its legitimacy. There are two parts to this problem. One is the vested interest of heavily involved parties such as the World Bank that questions the legitimacy of its providence.

The other, and more insidious, is that that the seven principles of the Code systematically validate the "win-win" narrative and legitimises the deals as at an appropriate response to the food security concerns of the investing countries. It does this in three ways:

1. by failing to address the drivers behind the deals in the first place and focussing on the deals as the valid solution to the investor countries' food security problems
2. by actively seeking to facilitate the investments, which it does by providing a framework for the negotiation of such deals.
3. by eliminating the concerns posed by the competing narrative which it does by asserting that the exploitative nature of the deals is able to be mitigated (simply) by adherence to each of the seven principles behind the Code.

Together, these effectively de-legitimize and elide the entire alternative discourse that is trying to expose and confront the potential for neo-colonial exploitation inherent in the deals. In addition, by eliding the competing narrative and its calls for alternative solutions, the international organisations are excluding the possibility of those alternative solutions.

The fact that the Code and its proponents are engaged in legitimizing one narrative over another, instead of acting to respond to the broader set of risks the deals present to both parties (that we examined previously and are basically ignored by the Code), brings the legitimacy of the Code itself into question.

I contend that the international organisations' response to the deals is a reactive response to quell the significant negative publicity generated by the anti-exploitation discourse opposing deals than a truly meaningful intervention against the risks the deals pose. This contention is supported both by the tokenistic choice of a non-binding instrument as the response mechanism as well as by the focus on the critical narrative of the deals instead of the problems with the deals themselves: that is, the forces behind them and the risks presented by them.

Moreover, the multilateral organisations' validation of the "win-win" narrative indicates their inherent support for the powerful investors, often who are important funders of the institutions (or are part of the institutions themselves), while superficially appearing to acknowledge the concerns of the vocal competing narrative and apparently addressing risks for the poor.

So to answer our earlier question on how these international organisations have co-operated to reconcile the conflicting sets of interests, focus on the forces behind the deals in the first place and attempt to mitigate the risks presented by the deals we can conclude quite simply that, while the Code remains their primary co-operative response, *they have not.*

Other Ways of Approaching the Problem?

The weaknesses of the *Code of Conduct* as the proposed co-operative international, multilateral response to the land deals speaks to the limitations of these international organisations in these circumstances. This suggests new institutions may be necessary and not just to respond to the deals themselves but to do more (and different things) to address the causes behind them in the first place, especially since these organisations are the very ones mandated to address global food insecurity. Whether these new institutions are additional organisations or simply finding new ways of doing things depends on how the interests of all parties, not just the investors, can be best served.

In this light, the recommendations of this paper are twofold. First, to undertake more critical analysis of the drivers behind the deals in the first place to elicit answers to questions such as: Do the situations in investor countries really warrant the need to gain control of foreign agricultural resources? And, what are the real drivers behind these policies and how else might these drivers be addressed?

Second, to research what, and how, alternative institutional arrangements can centralise the needs of the developing country agricultural communities in the context of foreign investors seeking reliable sources of staple foods. There are two parts to this exercise. One part is to undertake further research into alternative institutional mechanisms within the existing food security paradigm that consider the food security both of the investors and of the poorer communities in the developing countries. The research needs to consider the conditions in developing country agriculture, the opportunities for investment and the problems raised by the apprehension of countries like Saudi Arabia and China to rely totally on the international market to provide their food security needs, especially during times of crisis. An example of an alternative institutional mechanism in this regard might be regimes encouraging long-term bilateral arrangements that commit the investors to purchase agricultural production from developing country farmer groups at pre-agreed prices guaranteed by advanced partial payments. The combination of manageable cash-flows and known future income can enable the local economies to make the investments in productivity on their own terms without external interference. However, such arrangements still hold risks for the developing country: not least of which is the risk of crop failure and loss of expected income against which significant upfront investment has been made. Insurance (especially if funded by the purchaser) does offer the possibility of mitigating such risks. Nonetheless, this strategy does not easily protect the developing country farmers from exploitation by corporate industrialised agri-business moving in to take advantage of the opportunities presented by the contracts in the same way that the land investors can do directly under the terms of current land deal investment contracts. To address this would require concerted regulation—and enforcement—of corporate behaviour, both by the investor country and host country governments.

Other types of institutional mechanisms are also worth considering. Since World War II some more radical solutions to addressing international food security concerns have been proposed by the international organisations—especially the FAO—but never got off the ground due to the inability to reach consensus amongst the members. These include Boyd Orr's proposal in the 1950s for a World Food Board to manage a world food distribution policy based on the "economic advantages of a world food policy based on meeting human needs" (Shaw, 2007: 21) by improving the economic situation of all peoples and states, not just the to the economic benefit of the corporations controlling the industrialised food supply chain. This strategy was followed by ideas for an international commodity clearing house, a world food reserve and national food reserves in developing countries. These warrant contemporary re-visiting as possible models for alternative mechanisms to support the long term food needs of countries like Saudi Arabia and China as well as those of poorer developing countries.

The other part of this exercise is to research institutional mechanisms outside the food security paradigm, with the view of helping developing countries to effectively handle Foreign Direct Investment, whether in the agricultural sector or otherwise, and whether from countries pursuing food security or from profit-motivated corporations. An example of this already in place and directly relevant to exploitation of resources is the Extractive Industries Transparency Initiative (EITI). Although designed for extraction of mineral resources, it is potentially applicable to the extraction of agricultural resources. While the EITI is not perfect and has been the target of some of the same criticisms as the Code (Ocheje, 2006)—it too is an unenforceable legal instrument and it implicitly assumes resource extraction to be a good thing—but importantly, unlike the Code, it has legitimacy borne from the independence of the signatories from the profiteers from the resource extraction and the signatories have legislative powers over the extractors. In terms of having the legitimacy to help shape an emerging, positive norm it is seen at least by some independent analysis—including Ocheje—as "seeming to be on the right track" (Ocheje, 2006: 236). In these ways, the EITI offers some potential benefits over the Code for managing land deals which suggests that, at the very least, such mechanisms should be examined.

Another area of research outside the food security paradigm is the analysis—and objective assessment—of the potential value, and attendant risks, of economic protections by the developing countries around their agricultural (and more generally, FDI-targeted) sectors. While sudden export bans played a role in the price crisis of 2007 and 2008, more measured strategies may still be worthy of evaluation. Of course, a consideration must be that such strategies risk the opprobrium of the WTO and of conventional economic voices (largely speaking from, and in the interests of, the global North). Nonetheless, economic protectionism was hugely important in the rapid economic development of the so-called Asian Tigers like South Korea (see, for example, Chang, 2007), and continues to be standard policy in the agricultural sectors of the EU and USA, so there maybe lessons to be gained from further research in this area.

Conclusion

The proliferation of "food securing" land deals by countries like Saudi Arabia and China might be logical responses to address domestic anxieties about national food security, but there are risks posed by this strategy and they are reflective of a new set of security challenges for states. While risks and challenges for the hosts of the deals are articulated by the narrative of the deals as "neo-colonial" and "exploitative" of the developing countries, the risks are greater than these and include the possibility of the investor not securing their own food supplies as they desire and the possibility of increasing the propensity for political violence.

In response to the anti-exploitation narrative a legitimising discourse of the deals as "win-win" for both parties has developed, although there is little evidence to support the likelihood of "win-win" outcomes. The international organisations' response to the deals has bought into these competing discourses by legitimising the "win-win" narrative; attempting to invalidate the risks and concerns presented by the alternative narrative in its proposed solution, a Code of Conduct. Examining the Code in the light of these competing narratives has allowed us to identify a range of weaknesses and limitations for these international organisations in responding to the deals, in particular in that they are inadequate to reconcile the conflicting sets of interests, forget to focus the forces behind the deals in the first place and fail to act to mitigate the risks presented by the deals.

In a world of growing food insecurity, the demographic and environmental pressures being experienced by China and Saudi Arabia are not unique. Urbanisation, expanding middle classes and population growth are all demand-side forces also at work in other parts of Asia, Africa and Latin America. Together these point to what has been estimated by the FAO to require an increase in global food production capacity by some 70 percent by 2050 (FAO, 2009b) to meet the projected growth in global demand. At the same time, this needs to be achieved in the face of potential unpredictable pressures on production especially the likelihood of climate change impacts and what appears to be increasing soil erosion (as being explored by the GlobalSoilMap.net project, see Voosen, 2009) and water depletion (for example see USGS, 2010). This means that the drivers behind the land deals are not unique to the current investors and have implications for global food security and the future policies of other governments as well.

Thus to move forward in the interests of all parties and to establish sound precedents for a future that may hold more of the same, it is crucial that the adversarial—and uncooperative—responses entrenched in the competing narratives are overcome. This means attention needs to be is shifted away from the privileging and legitimising of the "win-win" narrative and that the concerns articulated in the competing narrative are acknowledged as valid. Furthermore, the broader scope of risks need to be recognized and long term solutions developed which address the food security concerns of both the investing countries and the local communities in the developing country hosts.

Notes

1 For example by the Consultative Group on International Agricultural Research (CGIAR) and the Asia-Pacific Association of Agricultural Research Institutes (APAARI) such as in APAARI's mission statement (2010).

2 For example by the Cargill transnational food corporation such as in their Corporate Responsibility statement (2010).

3 The Right to Food is codified in the Universal Declaration of Human Rights (1946, Article 25) and the International Covenant on Social, Cultural and Economic Rights (1966, Article 11).

4 For example by the World Health Organisation, in its Strategic Objectives (WHO, 2009).

5 For example, the Australian Government's bipartisan support for developing country food security policies that privilege international free trade over other forms of realising surety in local food supplies. These policies clearly support Australia's vested interest as a major agricultural exporter. See for example policy statements under the Rudd (Brown et al. 2008), Howard (Freeman et al. 2000), Keating (DFAT, 1996) and Hawke (Mauldon, 1986) governments.

6 The problem with urbanisation and food security which is not unique to Saudi Arabia, is that impoverished urban households often lack the means of subsistence that may be available to rural households. This means that food insecurity has the potential to be worse in cities than in rural areas, despite poverty—and unemployment—often being worse in the countryside. (e.g. Macours and Swinnen, 2008).
7 Saudi had an official unemployment rate in 2009 of 10.5 percent (Ministry of Economy and Planning cited in Baxter, 2010) 43 percent of which are aged between 20–24, Many non-official estimates of Saudi unemployment are considerably higher—up to 25 percent (Kawach, 2009).
8 For the complete assessment of the Saudi Arabia food policy situation, refer to this author's paper *Above Carrying Capacity: Saudi Arabia's External Policies for Securing Food Supplies* World Congress of Middle Eastern Studies, (Shepherd, forthcoming).
9 According to FAOSTAT (2010) database, Sudan's total arable and productive land is 19.5 million hectares.
10 The World Bank has since published a very detailed, and much anticipated, report into the so-called land grabs (World Bank, 2010) which has done nothing to change the situation. The report simultaneously acknowledges the risks to the poor while legitimising the deals as an appropriate strategy for investors and continues to espouse the proposed Code of Conduct as the solution to these issues.
11 The lack of jurisdiction of foreign entities in bilateral disputes in exemplified by the *Calvo Doctrine* and its effective absorption into the UN Charter of Economic Rights and Duties of States as an international norm (Sornarajah, 2004: 140–42).
12 Angola's dues of around US$12,000 are one thousandth of China's approx. US$12 million. Similarly, Saudi Arabia's approx US$3.5 million is compared to Zambia or Mozambique's US$4,000 contribution (see, FAO, 2010a).

References

Abdullah, S. 2010. Price Hike Hits Saudi Families Hard. *Arab News* April 13, 2010. Available online: http://arabnews.com/aaudiarabia/article42472.ece (Accessed 14 April 2010).

Al Arabiya 2009. Poverty in Saudi Arabia. Available online: http://www.youtube.com/watch?V=Ovprykaflv8 (Accessed 21 April 2010).

Apaari 2010. Apaari Misison Statement. Available online: http://www.apaari.org/about/ (Accessed 13 April 2010).

Baille, D. 2003. Tanzanian Parliament Blocks Government on GM Seeds. *Scidev. Net.* Available online: http://www.scidev.net/en/news/Tanzanian-Parliament-Blocks-Government-On-Gm-Seeds.html (Accessed 15 May 2010).

Baxter, E. 2010. Saudi Unemployment Rises to 10.5%. *Arabian Business*. Wednesday 21 April 2010.

BBC 2007. Mexicans Stage Tortilla Protest. *BBC News.* 1 February 2007. Available online: http://news.bbc.co.uk/2/hi/6319093.stm (Accessed 15 September 2009).

BBC 2008a. Food Riots Turn Deadly In Haiti. *BBC News.* 5 April 2008. Available online: http://news.bbc.co.uk/2/hi/americas/7331921.stm (Accessed 20 October 2009).

BBC 2008b. India Introduces Rice Export Ban. *BBC News.* 1 April 2008. Available online: http://news.bbc.co.uk/2/hi/7323713.stm (Accessed 20 October 2009).

BBC 2009a. Madagascar in Crisis. *BBC News.* 18 March 2009. Available online: http://news.bbc.co.uk/2/hi/africa/7935682.stm (Accessed 21 October 2009).

BBC 2009b. Madagascar Leader Axes Land Deal. *BBC News.* 19 March 2009. Available online: http://news.bbc.co.uk/2/hi/africa/7952628.stm (Accessed 22 October 2009).

Bhalla, S. 2007. Second Among Equals: The Middle Class Kingdoms of India and China. *Peterson Institute For International Economics.* Available online: http://www.oxusinvestments.com/files/pdf/ne20090106.pdf (Accessed 4 February 2010).

Blas, J. 2008. Foreign Fields: Rich States Look Beyond Their Borders for Fertile Soil. *Financial Times.* 19 August 2008.

Brown, L. 2008a. A Threat to Global Stability. *Global Asia*, Vol. 3, No. 2, Summer 2008, 10–14.

Brown, N. 2008b. *High Food Prices, Food Security and the International Trading System.* Informa National Food Pricing Summit, Sydney, 29–30 September 2008.

Cargill 2010. *Corporate Responsibility.* Available online: http://www.cargill.com/corporate-responsibility/pov/food-security/index.jsp (Accessed 12 April 2010).

Chanboreth, E. and Hach, S. 2008. *Aid Effectiveness in Cambodia.* Brookings Institution: The Wolfensohn Center For Development Working Papers No. 7. Available online: http://www.brookings.edu/papers/2008/12_cambodia_aid_chanboreth.aspx (Accessed 24 May 2010).

Chang, H-J. 2007 *Bad Samaritans: Rich Nations, Poor Policies and the Threats to the Developing World.* London. Random House Business Books.

Chapman, P. 2008. *Bananas: How He United Fruit Company Shaped the World.* Edinburgh: Canongate.

China Daily 2008. Firms Will Grow Rice in Africa. *China Daily.* 9 May 2008. Available online: http://farmlandgrab.org/2346 (Accessed 18 May 2010).

China Post 2008. Vietnam Lifts Rice Export Ban on News of Bumper Harvest. *The China Post.* Available online: http://www.chinapost.com.tw/business/asia/vietnam/2008/06/21/162003/Vietnam-Lifts.htm (Accessed 30 March 2010).

Davis, M. 2001. *Late Victorian Holocausts: El Nino Famines and the Making of the Third World.* London: Verso.

DFAT 1996. *Food Security and Trade: A Future Perspective.* Canberra: Department of Foreign Affairs and Trade.

Emerging Asset Management 2010. *African Land Fund.* Available online: http://www.eaml.net/?func=pageafricanlandfund (Accessed 12 February 2010).

England, A. and Blas, J. 2009. Qatar Swf to Target Food and Energy. *Financial Times.* Available online: http://farmlandgrab.org/2827 (Accessed 18 May 2010).

FAO 2002. Net Trade in Food. Available online: http://faostat.fao.org/portals/_faostat/documents/pdf/map05.pdf (Accessed 26 February 2010).

FAO 2009a. Countries in Crisis Requiring External Assistance. *Food and Agriculture Organisation: Global Information And Early Warning System*. Available online: http://www.fao.org/giews/english/hotspots/index.htm (Accessed 13 August 2009).

FAO 2009b. How to Feed the World by 2050. *Food and Agriculture Organisation of the United Nations.* Available online: http://www.fao.org/news/story/en/item/35571/icode/ (Accessed 14 May 2010).

FAO 2009c. *One Sixth of Humanity Undernourished – More Than Ever Before.* Available online: http://www.fao.org/news/story/en/item/20568/icode/ (Accessed 22 August 2009).

FAO 2009d. *The State of Food Insecurity in the World.* Rome: United Nations Food and Agriculture Organisation.

FAO 2009e. World Summit on Food Security. Rome, 16–18 November 2009. Available online: http://www.fao.org/wsfs/wsfs-addresses/wsfs-mondayam/en/ (Accessed 22 May 2010).

FAO 2010a. *Regular Programme Contributions*. Available online: http://www.fao.org/about/18791-0-0.pdf (Accessed 14 May 2010).

FAO 2010b. *Regular Programme Contributions: Top 25 Contributors*. Available online: http://www.fao.org/about/18792-0-0.pdf (Accessed 14 May 2010).

FAOSTAT 2010. FAO Statistical Database. Available online: http://faostat.fao.org/default.aspx (Accessed 22 March 2010).

FIAN 2010. Stop Land Grabbing Immediately! Available online: http://www.fian.org/cases/letter-campaigns/stop-land-grabbing-immediately/?searchterm=land%20deals (Accessed 25 May 2010).

Finch 2010. Oz-Based Fund Eyes $600m in Agro-Tech. *The Phnom Penh Post.* Available online: http://farmlandgrab.org/10585 (Accessed 25 May 2010).

Freeman, F., Melanie, J., Roberts, I., Vanzetti, D., Tielu, A. and Beutre, B. 2000. *The Impact of Agricultural Trade Liberalisation on Developing Countries.* Canberra: Australian Bureau of Agricultural and Resource Economics.

Galtung, J. 1969. Violence, Peace and Peace Research. *Journal of Peace Research*, Vol. 6, No. 3, 167–91.

GRAIN 2008a. Food Crisis and the Hybrid Rice Surge. *Grain.Org*. Available online: http://www.grain.org/seedling/?id=551 (Accessed 18 May 2010).

GRAIN 2008b. Grain Briefing. *Grain.Org*. October 2008. Available online: www.grain.org/go/landgrab (Accessed 24 February 2009).

GRAIN 2008c. Grain Briefing Annex. *Grain.Org*. October 2008. Available online: www.grain.org/go/landgrab (Accessed 24 February 2009).

GRAIN 2009a. Farmlandgrab.org. Available online: www.farmlandgrab.org (Accessed 12 February 2010).

GRAIN 2009b. Rice Land Grabs Undermine Food Sovereignty, in *Africa Against the Grain*. January 2009. Available online: http://www.grain.org/articles/?id=46 (Accessed 15 May 2010).

GRAIN 2010. Donors Open the Door for Land Grab In PNG. Available online: http://farmlandgrab.org/12716 (Accessed 25 May 2010).

Gurr, T.R. 1970. *Why Men Rebel*. Princeton: Princeton University Press.

Guyler, J. and Loney, J. 2008. Haiti's Government Falls After Food Riots. *Reuters*. Saturday 12 April 2008. Available online: http://www.reuters.com/article/topnews/idusn1228245020080413 (Accessed 19 September 2009).

Hoering, U. 2009. Hauptdestination Afrika: Die Neue Landnahme. *Informationsbrief Weltwirtschaft und Entwicklung*. Available online: http://www.weltwirtschaft-und-entwicklung.org/wearchiv/042ae69b9c0b33d01.php (Accessed 25 May 2010).

Homer-Dixon 1991. On the Threshold: Environmental Changes as Causes of Acute Conflict. *International Security*, Vol. 16, No. 2 (Autumn), 76–116.

Homer-Dixon 1994. Environmental Scarcities and Violent Conflicts: Evidence from Cases. *International Security,* Vol. 19, No. 1 (Summer) 5–40.

Horta, L. 2008. The Zambezi Valley: China's First Agricultural Colony? *Africa Policy Forum*. 9 June 2008. Available online: http://forums.csis.org/africa/?p=120 (Accessed 6 August 2009)

Irin 2009. Africa: Tractored Out by "Land Grabs"? *IRIN Project of UN Office for Coordination of Humanitarian Affairs*. Available online: http://www.irinnews.org/report.aspx?reportid=84320 (Accessed 25 May 2010).

Kawach, N. 2009. Saudi Arabia Seeks Private Sector's Help to Create Jobs. *Emirates Business*. Sunday, 12 July 2009.

Kielburger, C. 2009. Food Deals Leave Developing Nations Hungry. *Toronto Star*. Available online: http://www.thestar.com/news/globalvoices/article/742765--food-deals-leaving-developing-nations-hungry (Accessed 23 April 2010).

Kirby, W. 2009. China's Emerging Middle Class: Beyond Economic Transformation. *Brookings Institution*. Available online: http://www.brookings.edu/~/media/files/events/2009/0922_0923_China/20090922_China_Middle_Class_1.pdf (Accessed 23 February 2010).

KSA MOFA 2010. Custodian of the Two Holy Mosques Receives Minister of Commerce and Industry. Kingdom Of Saudi Arabia Ministry Of Foreign Affairs. Available online: http://www.mofa.gov.sa/detail.asp?insectionid=3981&innewsitemid=88796 (Accessed 15 May 2010).

Lavallee, G. 2009. US Eyes Future Oil Deals in Southern Sudan. *Middle East online*. Available online: http://www.middle-east-online.com/english/?id=30475 (Accessed 26 May 2010).

Licadho 2009. Land Grabbing and Poverty in Cambodia: The Myth of Development. Available online: http://www.licadho-cambodia.org/reports/files/134licadhoreportmythofdevelopment2009eng.pdf (Accessed 25 May 2010).

Macours, K. and Swinnen, J. 2008. Rural-Urban Poverty: Differences in Transition Countries. *World Development*, Vol. 36, No. 11, 2170–87.

Madoffe, S. 2009. Africa: Biofuels and Neo-Colonialism. *Weekly Forum for Social Justice in Africa.* Available online: http://allafrica.com/stories/200906040880.html (Accessed 22 May 2010).

Mauldon, R. 1986. Food Trade and Food Security in ASEAN and Australia. *Policy Issues in ASEAN/Australian Trade*, Vol. 1986, 250–63.

Meinzen-Dick, R. and Markelova, H. 2009. Necessary Nuance: Towards a Code of Conduct in Foreign Land Deals, in Kugelman, M. and Levenstein, S. (eds) *Land Grab? The Race for the World's Farmland.* Washington, DC: Woodrow Wilson International Centre for Scholars.

Mittal, A. 2010. (Mis)Investment in Agriculture: The Role of the International Finance Corporation in the Global Land Grab. *The Oakland Institute.* Available online: http://www.oaklandinstitute.org/?q=node/view/555 (Accessed 25 May 2010).

Monbiot, G. 2008. Manufactured Famine. *The Guardian.* 26 August 2008.

Niblock, T. 2006. *Saudi Arabia: Power, Legitimacy and Survival.* Abingdon: Routledge.

Nicoll, A. 2008. The Simmering Food Crisis. *Iiss: Strategic Comments.* London: The International Institute For Strategic Studies.

Nkemdirim, B. 1977. Reflections on Political Conflict, Rebellion and Revolution in Africa. *Journal Of Modern African Studies*, Vol. 15, No. 1, 75–90.

Nkrumah, K. 1965. *Neo-Colonialism: The Last Stage of Imperialism* London: Thomas Nelson & Sons.

Ocheje, P. 2006. The Extractive Industries Transparency Initiative: Voluntary Codes of Conduct, Poverty and Accountability in Africa. *Journal Of Sustainable Development in Africa*, Fall 2006, 222–39. Available online: http://www.jsd-africa.com/jsda/fall2006/pdf/arc_The%20extractive%20industries%20transparency%20initiative.pdf (Accessed 25 May 2010).

Okruhlik, G. 1999. Rentier Wealth, Unruly Law and the Rise of Opposition: The Political Economy of Oil States. *Comparative Politics*, Vol. 31, No. 3, 295–315.

Oryza. 2008. Bangladesh Bans Rice Exports. *Oryza online.* Available online: http://oryza.com/asia-pacific/bangladesh-market/8469.html (Accessed 30 March 2010).

Parsons, C. 2009. Congo Hopes to Finalize S. Africa Land Deal in 2009. *Reuters*. Available online: http://af.reuters.com/article/topnews/idafjoe58m00m20090923 (Accessed 22 April 2010).

People's Republic of China. 2008. National Food Security and Long-Term Planning Framework, (Trans.) Professor Zhang Qing Chinese Academy Of Social Sciences. Beijing: Available online: http://www.gov.cn/jrzg/2008/-11/13/content_1148414.ttm (Accessed 12 March 2010).

Petras, J. 2008. The Great Land Giveaway: Neo-Colonialism by Invitation. *Centre for Research on Globalisation.* Available online: http://www.globalresearch.ca/index.php?context=va&Aid=11231 (Accessed 15 December 2008).

Philippine Government. 2009. Saudi Firms See RP as Food Production Hub in the Far East. *Philippine Government Portal.* Available online: http://www.gov.ph/index.php?option=com_content&task=view&id=2001705&itemid=2 (Accessed 23 April 2010).

Raine, S. 2009. *China's African Challenges.* Abingdon: Routledge.

RCCI. 2008. Saudi Importers May File Suit Against Indian Rice Exporters. *Riyadh Chamber of Commerce and Industry.* Available online: http://www.riyadhchamber.com/newsdisplay.php?id=631 (Accessed 14 April 2010).

Reuters. 2008a. Kuwait Eyes Asia for Food Supply, Agriculture. *Reuters.* Available online: http://farmlandgrab.org/2457 (Accessed 18 May 2010).

Reuters. 2008b. Vietnam Lifts Rice Export Ban, Manila Seals Deal. *Reuters India.* Available online http://in.reuters.com/article/domesticnews/idinsp738702008061 8?Sp=True (Accessed 30 March 2010).

Reuters. 2009. Kuwait Eyes African Investments. *Reuters.* Available online: http://www.tradearabia.com/news/newsdetails.asp?sn=agri&artid=164576 (Accessed 22 May 2010).

Saudi Gazette. 2010. New Sr20b Agriculture Fund to Offer Easy Loans. *Saudi Gazette International Edition.* Friday, 16 April 2010. Available online: http://www.saudigazette.com.sa/index.cfm?method=home.regcon&contentid=2009012727717 (Accessed 16 April 2010).

Shaheen, A.R. 2009. Poverty Alleviation Becomes a Priority for Saudi Arabia. *Gulf News.* Available online: http://gulfnews.com/news/gulf/saudi-arabia/poverty-alleviation-becomes-priority-for-saudi-arabia-1.540390 (Accessed 21 April 2010).

Shaw, D.J. 2007. *World Food Security: A History Since 1945.* Baskingstoke: Palgrave Macmillan.

Shepherd, B. Forthcoming. *Above Carrying Capacity: Saudi Arabia's External Policies for Securing Food Supplies.*

Socheat, A. 2010. Australian Agro-Deal in Cambodia Carries Risks, Rewards. *Deutsche Presse.* Available online: http://farmlandgrab.org/10750 (Accessed 25 May 2010).

Sornarajah, M. 2004. *International Law on Foreign Investment.* Cambridge: Cambridge University Press.

Speildoch, A. and Murphy, S. 2009. Agricultural Land Acquisitions: Implications for Food Security and Poverty Alleviation, in Kugelman, M. and Levenstein, S. (eds) *Land Grab? The Race for the World's Farmland.* Washington, DC: Woodrow Wilson International Centre for Scholars.

Steinberg, S. 2008. Financial Speculators Reap Profits from Global Hunger. Centre For Research On Globalization. Available online: http://globalresearch.ca/index.php?context=va&aid=8794 (Accessed 14 April 2010).

Sudan Tribune. 2009. New York Investment Firm Mulling More Land Leases in S. Sudan. *Sudan Tribune.* Available online: http://www.sudantribune.com/spip.php?article30891 (Accessed 25 May 2010).

Sudan Tribune. 2010. Chinese Firm Given Land Deal in Sudan. *Sudan Tribune*. Available online: http://www.sudantribune.com/spip.php?article34444 (Accessed 22 April 2010).

Taylor, J. and Buncombe, A. 2008. The Food Crisis Begins to Bite. *The Independent*. 25 April 2008.

The Economist. 2009. Buying Farmland Abroad: Outsourcing's Third Wave. *The Economist*. 21 May 2009.

UK Zambians. 2010. Saudi Investment Mission Visiting Zambia. UK Zambians. Available online: http://ukzambians.co.uk/index.php?option=com_content&task=view&id=3129&itemid=533 (Accessed 23 April 2010).

UNEP. 2005. Saudi Arabia: Urban Population. Globalis Interactive World Development Database (UN University and United Nations Environment Program). Available online: http://globalis.gvu.unu.edu/indicator_detail.cfm?indicatorid=30&country=sa (Accessed 22 April 2010).

US Bureau of the Census. 2010. International Database. Available online: http://www.census.gov/ppc/www/idb/informationgateway.php (Accessed 13 March 2010).

USGS. 2010. Groundwater Depletion. United States Geological Survey. Available online: http://ga.water.usgs.gov/edu/gwdepletion.html (Accessed 18 May 2010).

Vezirgiannidou, S.-E. 2010. *Entering the Zone of Agreement: The United States on Climate Change Negotiations*. Cambridge: Cambridge University Press.

Voosen, P. 2009. Bold Soil-Mapping Venture Seen as Crucial to Efforts on Climate, Agriculture. *The New York Times*. Available online: http://www.nytimes.com/gwire/2009/08/07/07greenwire-bold-soil-mapping-venture-seen-as-crucial-to-e-74103.html (Accessed 18 May 2010).

Walid, T. 2009. Qatari Swf Food Unit Eyes Pava Stake, Seals Sudan Deal. *Reuters*. Available online: http://af.reuters.com/article/investingnews/idafjoe59s0lw20091029?pagenumber=1&virtualbrandchannel=0 (Accessed 22 May 2010).

Weatherly, R. 2006. *Politics in China Since 1947: Legitimizing Authoritarian Rule*. Oxford: Routledge.

WeltHungerHilfe 2010. Brennenpunkt 8: Land Grabbing – Den Armen Wird Der Boden Unter Den Füßen Weggenzogen. Available online: http://www.welthungerhilfe.de/positionspapiere.html?&0= (Accessed 25 May 2010).

WFP. 2009. *Operations and Resourcing Update*. July 2009 to July 2010. Available online: http://documents.wfp.org/stellent/groups/public/documents/op_reports/wfp110572.pdf (Accessed 22 May 2010).

WHO. 2009. Strategic Objectives 9 Workplan 2008–2009. Available online: http://www.euro.who.int/nutrition/20090106_1 (Accessed 12 May 2010).

World Bank, FAO, IFAD and UNCTAD. 2010. Principles for Responsible Agricultural Investment that Respects Rights, Livelihoods and Resources. Extended Version, 25 January 2010.

World Bank. 2010. Rising Global Interest in Farmlands: Can it Yield Sustainable Results? 7 September 2010. Available online: http://siteresources.worldbank.

org/INTARD/resources/ESW_Sept7_final_final.pdf (Accessed 11 September 2010).

Zamora, S. 1999. Econmic Relations and Development: 2. Codes of Conduct, in Joyner, C. (ed.) *The United Nations and International Law*. Cambridge: Cambrige University Press.

Conclusions
Securing New Security Frontiers

Daniel S. Papp

As the definition of "frontier" in the *Merriam-Webster Online Dictionary* makes clear, *New Security Frontiers* is indeed an appropriate lead title for this volume and the chapters it contains. According to the dictionary, a frontier is, among other things, both "a line of division between different or opposed things" and "a new field for exploitative and developmental activity."[1]

Under these definitions, the world has already entered a period during which, in the field of security, "a line of division between … opposed things" has been crossed. Specifically, that line is between old ways of thinking about security, focusing primarily and sometimes exclusively on military affairs, and new ways of thinking about security, in which more comprehensive ways of thinking about and planning for security are being adopted. Beyond the constant dangers posed by terrorism and other obvious causes of warfare, shortfalls of energy resources, scarcity of non-fuel mineral resources, shortages of food, lack of fresh water, encroachment of desertification, and dangers posed by cyber attacks are increasingly recognized as presenting as much, and over time perhaps even more, of a challenge and threat to security as do guns, bombs, and missiles. Some of the security challenges and threats addressed in this volume are widely perceived as imminent, while others present challenges and threats in the mid- to long-term.

At the same time, as we cross this line of division between thinking about "old" security challenges and threats and "new" security challenges and threats, new ways of dealing with these challenges and threats must be adopted. Thus, as the dictionary's second definition of "frontier" makes clear, "a new field for exploitative and developmental activity" in security thinking is opening up. One such approach is a recognition that the "Three D's" approach to security –diplomacy, development, and defense – adopted to an extent by the Bush administration in 2001, is a clear step forward in security thinking despite its perceived shortcomings. An even newer approach to thinking about security is the "whole-of-government" approach adopted by the Obama administration in 2010. Both these approaches to security will be discussed later in this chapter.

The preceding chapters in this volume, four of which focus on energy, one on non-fuel mineral resources, one on water, and one on food, make it clear that the international community in the early twenty-first century is confronted by a combination of interconnected challenges and threats to security that can not be resolved by military means alone. As a result, although the dangers of superpower confrontation, nuclear annihilation, and main force warfare have receded into the background, the early twenty-first century world is a world where national and global security are at risk.

Emerging Challenges and Threats to National and Global Security

At the outset of this chapter, each of the emerging challenges and threats to national and global security listed above – shortfalls of energy resources, scarcity of non-fuel mineral resources, lack of fresh water, and shortages of food – require additional words of concluding commentary. In this concluding chapter, two additional related areas of security frontiers – encroachment of desertification and threats posed by cyber attacks – will also be briefly explored.

Energy Shortfalls

Security-related concerns over energy are not new. During the twentieth century, such concerns on the part of industrialized nations including the Unites States date back at least as far as World War II, generated by a combination of fears about access to, availability of, control over, denial of, prices of, competition over, and depletion of energy resources.

During the last years of the twentieth century and the first years of the twenty-first, concerns about possible global energy shortfalls, for the most part oil, heightened for many reasons. During the 1980s, concerns about the ability of the Organization of Petroleum Exporting Countries (OPEC) to determine oil production levels and prices created consternation in many Western industrialized states. As the 1990s opened, Saddam Hussein invaded Kuwait in a transparent effort to have Iraq become the so-called "swing producer" of oil on a global basis, that is, the producer that could by itself determine global oil prices, if not production levels. As Operation Desert Storm expelled Iraq from Kuwait, Hussein turned to a denial strategy, setting Kuwaiti and Iraqi oil wells ablaze. Meanwhile, throughout the last decade of the twentieth century and first decade of the twenty-first century, China and India's energy consumption soared, driving demand and prices upward and creating concerns about future conflict over and shortage of energy resources. In the first few years of the twenty-first century, natural and man-made disasters ranging from hurricanes and tsunamis to poorly constructed and improperly operated off-shore oil production facilities also disrupted global oil production, as did political unrest which swept the Middle East.

Concerns about energy shortfalls are nothing new in the United States. For example, the American energy crises of the early 1970s, fueled primarily by concerns about access to, prices of, and dependence on Middle Eastern oil, led the United States to improve energy use efficiency, expand energy resource production, and develop alternative energy resources. Some efforts were more successful than others. Nevertheless, the harsh reality is that while the growth of American energy use slowed, overall American energy consumption expanded during the late twentieth and early twenty-first centuries.[2]

The United States was not the only country concerned about energy shortfalls in the late twentieth and early twenty-first centuries. Europe and Japan shared many of the same American concerns, and adopted many of the same American responses with somewhat the same measure of success. Although European countries and Japan used less energy on a per capita basis than the United States, per capita energy use in Europe and Japan also climbed. Many European states were also uncomfortable that a large percentage of their oil and natural gas imports came from Russia and Central Asian states, given especially Russia's seeming willingness on occasion to use oil and gas exports to achieve political ends.[3]

Also, as note above, new demands were placed on the world's energy supplies in the late twentieth and early twenty-first century as China and India enjoyed rapid and sustained economic growth. This raised the specters both of increased rivalry for energy resources and higher prices for energy resources. China by itself as of 2006 was the world's second largest consumer and third largest producer of primary energy, accounting for half of the growth in the world's demand for energy between 2000 and 2005 alone. Largely self-sufficient in coal, China nevertheless imports up to half of its petroleum needs.[4] China thus joined the United States, Japan, and European states as a location where energy security is a paramount security concern.

By the opening of the second decade of the twenty-first century, then, energy security had become a global issue. This concern took many forms, including access to, availability of, control over, denial of, prices of, competition over, depletion of, and new methods of providing energy resources. It is beyond debate that energy security is one part of the twenty-first century's new security frontier.

Non-Fuel Mineral Resources Scarcity

Much the same is true of non-fuel mineral resources. Throughout the industrial age, access to, availability of, control over, and depletion of non-fuel mineral resources, also called strategic mineral resource, have waxed and waned as security concerns. For example, the 1870–71 Franco-Prussian War was fought at least in part because of Germany's desire to control the iron ore resources of Alsace-Lorraine, and many of the colonial disputes of the late nineteenth and early twentieth century were driven by efforts to control mineral resources. More recently, during World War II, the policies of most of the major combatant states were strongly influenced by strategic mineral resource concerns. Indeed, a strong case can be made that Japan initiated the creation of the so-called "East Asian Co-Prosperity Sphere" more for reasons of control of and guaranteed access to strategic resources than for reasons of national pride.

More recently, in the 1970s and 1980s, a number of national security analysts argued that Soviet support for national liberation movements in Angola, Mozambique, Namibia, South Africa, Zaire, and elsewhere was actually a clever long-term strategy designed to cut the West off from one of its primary sources of non-fuel mineral resources such as chromite, cobalt, diamonds, gold, platinum, and titanium.[5] While the dissolution of the Soviet Union removed these concerns, the fact remains that southern Africa into the twenty-first century continued to produce a large percentage of the world's most critical non-fuel mineral resources.[6]

Concerns over non-fuel mineral resource shortages dissipated in the late twentieth century after the Soviet Union collapsed and the rich mineral resources of Russia and other former Soviet republics became available on the global market. However, they did not totally disappear, and the early twenty-first century has in some quarters seen a resurgence of such concerns, although for the most part they remain muted in terms of national security. For example, the announcement in 2002 of a cooperative US Geological Survey (USGS) project to undertake global assessments of undiscovered deposits of copper, gold, lead, nickel, platinum-group metals, potassium, and zinc (the Global Mineral Resource Assessment Project) began by observing:

> In today's global economy, a nation's economic security depends on its access to adequate mineral resource supplies from a variety of domestic and international sources. Although no global shortages of non-fuel mineral resources are expected in the near future, the growing demand for and utilization of mineral resources require continued exploration and development of as-yet undiscovered mineral deposits. However, the ability to find and develop new mineral resources is increasingly impacted by competing land uses and values, concerns about the environmental effects of mining, and the world's growing population.[7]

Other recent discussions and studies have also noted the link between strategic minerals and national security. One of the most insightful is the World Foresight Forum's *Rare Earth Elements and Strategic Mineral Policy*, which details the evolution and development of, as well as the forces behind, strategic minerals policies of the United States, the United Kingdom, and Japan, and declares that:

> Ensuring and safeguarding access to rare earth elements and other strategic mineral resources is quickly emerging as a strategic policy priority and a number of states are designing and implementing new policies aimed at increasing material security. … [T]his report provides an insight into what drives policies on strategic non-fuel mineral resources.[8]

Indeed, occasional publications even make the claim that the presence of US military forces in Afghanistan is driven as much by the claimed existence of up to a trillion dollars worth of mineral resources in Afghanistan as by the war on terrorism![9]

Be that as it may, while no credible sources claim that the global community faces an impending shortage of strategic mineral resources, many recognize that temporary shortfalls are possible, and that curtailing mineral deliveries can be used as a tool of political pressure such as when in 2010 China temporarily froze rare earths exports to Japan in retaliation for Japan's capture of a Chinese vessel near the disputed Senkaku islands in the East China Sea.[10] Clearly, then, although scarcity of strategic mineral resources and the use of strategic minerals as a policy tool is rarely discussed in public forums, it is much on the minds of national policy-makers.

Lack of Fresh Water

Conflict over fresh water is not new. The Pacific Institute for Studies in Development, Environment, and Security's "Water Conflict Chronology" lists literally thousands of instances of such conflict dating back to the time of the ancient Sumerians up until today.[11] However, there is a growing global awareness of just how serious a challenge and threat conflict over fresh water presents. Indeed, all evidence available today suggests that such conflict will continue and quite probably worsen.

The earth contains an estimated 1,386 billion cubic kilometers of water, but approximately 97.5 percent of it is salt water, unfit for human consumption.[12] Given the combination of increased global population, expanding environmental degradation, continuing and accelerating industrial growth, and advancing desertification (discussed below), it is not surprising that an article by two researchers at Sandia National Laboratory in a 2008 edition of *Nature* magazine devoted to "Water: Under Pressure" observed that by 2025, half the nations in the world will face shortages of or stress on fresh water supplies, and by 2050, as much as 75 percent of the world's population could face fresh water sshortages.[13] As the article observed:

> drinking water supplies, agriculture, energy production and generation, mining and industry all require large quantities of water. In the future, these sectors will be competing for increasingly limited freshwater resources, making water supply availability a major economic driver in the twenty-first century.[14]

Today, the Middle East, Africa, and certain regions in South America are all locations where conflict over fresh water may erupt. The waters of ancient water systems such as the Nile, Tigris, and Euphrates Rivers are all under increasing stress. According to some estimates, as many as 50 countries on five continents, including the world's two most populous states, China and India, face impending fresh water shortages.[15]

Even in the United States, disagreements over water usage are frequent, especially in the American West.[16] But such disagreements are not limited to the American West. For example, water use disagreements also make front page news in the south, witness the continuing dispute between Georgia, Alabama and Florida over water use from the Chattahoochee River.[17] Fortunately, however, disputes in the US over water usage have not erupted into violent conflict.

Can anything be done to alleviate the possibility of expanded conflict over fresh water? At least two sets of solutions may be possible, one via policy and the other via technology. On the policy front, international agreements on how to share reservoirs, rivers, lakes, and other sources of fresh water must be elevated on the agenda of the global community, and quickly. On the technological front, emphasis on research on advanced membrane separation technologies and desalinization must be expanded, as must the development of technologies for purification and use of non-traditional fresh water sources such as wastewater, brackish groundwater, and other gray water sources.

The challenge is immense, and so too are the threats. But there is no doubt that the growing shortage of fresh water on both a local and a global scale are both challenges and threats to local, national, regional, and global security.

Encroachment of Desertification

Closely related to the security challenges and threats presented by the growing shortage of fresh water is the increasing encroachment of desertification in many areas of the world. The UN Convention to Combat Desertification (UNCCD) defines desertification as "land degradation in arid, semiarid, and dry sub-humid areas resulting from various factors, including climatic variations and human activities," with land degradation in turn being defined as the reduction or loss of the biological or economic productivity of dry-lands. Dry-lands by themselves occupy 41 percent of the earth's land surface, and are home to about 2 billion people, that is, one third of all humanity.[18]

Desertification leads directly to population migration, additional pressure on available scarce resources including food and water, and on occasion, conflict and warfare. Indeed, in the Darfur region of Sudan, desertification is one of the primary factors behind population migration and conflict. Multiple inputs contribute to this ongoing human tragedy, which may have as of 2011 claimed as many as half a million lives. Since about 1970, deserts have spread southward in the region by an average of 100 kilometers per decade. Overgrazing has been a primary cause of this as the number of livestock has exploded from 27 million animals to about 135 million. Deforestation has also been a primary contributor as Sudan lost as much as 12 percent of its forests in only 15 years. Nature has also played a role in the crisis as northern Darfur's precipitation has fallen by one third in the past 80 years.[19]

Desertification is occurring on every continent except Antarctica. Some of the more prominent locations are the Sahel in West Africa, the Gobi Desert west of Beijing, much of Mongolia, the Sahara across northern Africa, the Namib in southern Africa, the Atacama in South America, much of the high plains in the American West, and as already discussed, the Darfur region of Sudan.

As a security challenge and threat, desertification creates instability not only in regions and countries that experience desertification, but also, because of population migration, in regions and countries that border such areas. Thus, as the world moves into an era of new security frontiers, desertification is something that cannot be overlooked or ignored.

Food Shortages

Food shortages and famines have been causes of conflict throughout history. The holy scriptures of many religions tell stories of famine and conflict. For example, the Bible in Revelations 6:1–8 links conquest, famine, war, and death as the four horsemen of the apocalypse, and Buddhism's Bodhisattva Sutra declares, "May the aeon of disease, famine and war be pacified." More recently, in the early nineteenth century, Thomas Malthus predicted that population growth would outstrip the growth of food production, leading to famine and conflict.[20]

A combination of agricultural advances, the Green Revolution, opening new land areas to food production and other factors have prevented the on-set of the more extreme views of religious texts and Malthusian predictions. However, there is no denying that food shortages, often linked to shortages of fresh water, encroachment of desertification, increases of food prices, supply disruption, and natural disasters are major causes of conflict and war in the twenty-first century. In the first decade of the twenty-first century alone, over 30 countries, most in the developing world including Bangladesh, Burundi, Congo, Egypt, Haiti, Somalia, Sudan, Venezuela, and Zimbabwe have experienced riots, violent protests, or warfare linked to food shortages.[21]

For policy-makers and people in countries experiencing riots, violent protests, or warfare because of food shortages, the challenge and threat to national security is obviously immediate. How to alleviate food shortages and end conflict is thus an immediate concern.

For policy-makers and people in countries currently with sufficient food, shortages of food elsewhere may not be an immediate problem, but such shortages cannot be ignored as they present both policy and ethical problems. What policies should be initiated to quell food-shortage-induced violence and fighting in countries beset by food shortages? What should be done to alleviate food shortages so unrest and violence does not spill over international boundaries? Given all the other demands on resources available to governments, non-governmental organizations, and corporations, what are the proper policies to implement to ensure security from violence, conflict, and warfare induced by food shortages?

These are difficult questions, and the answers, if any, are extremely complex. There is little doubt, however, that food shortages on both a national and global level are part of the equation that must be answered if the twenty-first century's security frontiers are to be secured.

Cyber Attack Threats

Given the pervasive nature of information and communication technologies, the threat of cyber attacks must also be considered as a concern of new security frontiers. Regardless of whether the issue is delivery of energy and non-fuel mineral resources, the provision of water resources on a timely and needed basis, or the delivery of food, the chances are great that information and communication technologies are integrally involved. Thus, defending against cyber attacks is a critical segment of securing emerging security frontiers.

Several examples drive the point home. Aside from thousands of cases of identity theft, hackers among other things have recently:

1. gained access to computerized controls of Arizona's Salt River Project near Phoenix;
2. crashed a computer network at an Ohio nuclear plant;
3. altered voice messages and commands at the Ozarks Electric Company; and
4. launched a worm attack against American supervisory control and data acquisition systems that could have sabotaged operational systems.[22]

In addition, in 2009 and 2010, hackers penetrated the US power and electrical grid and left software programs that could disrupt the system.[23] Some national security experts are also concerned that hackers could take over a nuclear power plant, financial networks, key utilities, railroad transportation systems, and natural gas pipelines.[24]

While the danger from these attacks varied, the point is that cyber security has become part of the frontiers of the twenty-first century security equation. Challenges and threats to cyber systems carry with them at least three implications for security.[25]

First, because of modern society's dependence on information and communication technologies in almost all areas of human endeavor, the definition of security will expand. This is because:

1. information and communication technologies make the transfer of information easier as organizational boundaries, state borders, and other lines of demarcation within and between states and other sub-national, national, and international actors become increasingly permeable; and
2. as cyber technologies become increasingly pervasive and societies grow increasingly dependent upon them, an information society's vulnerability to alteration, disruption, or destruction of those technologies grows.

For example, in finance and banking, funds can be transferred electronically at a moment's notice from one location to another almost anywhere in the world. This decreases the ability of states to control and monitor financial flows and business transactions across borders. Whereas in the past information security was primarily a corporate or personal issue, the free flow of information has potential to raise information security issues to the national level. The transfer of funds, data, and other forms of information electronically across state boundaries opens opportunities for electronic theft, electronic blackmail, electronic corruption, electronic data alteration, and system disruption via electronic assault upon the economic, political, and social stability and well-being of a state. System disruption is clearly a security concern, and under certain conditions, electronic theft, electronic blackmail, electronic corruption, and electronic data alteration could be as well. Thus, more issues will be security issues.

Second, more diverse sources of challenges and threats to national security will emerge. These will emanate from sources that in the past may not have been national security challenges or threats. As information and communication technologies become less expensive and easier to use, they will be more widely adopted and employed. Their employment will no longer be limited to "leading edge" industries and organizations, nor will their employment be limited to select organizational functions and processes. More people, institutions, and organizations will have more access to information. Except for sensitive national and corporate data, this increased quantity of data will be accompanied by increased dissemination and access. Similarly, the locations at which information is located and from which information can be accessed will proliferate.

During the industrial era, it was a rare occurrence for a single individual to present a true threat to security. Unfortunately, not all of the users of advanced information and communication technologies will have the best interests of a given state in mind. Inevitably, attempts at electronic theft, blackmail, corruption, data alteration, and service disruption will occur. Some will be successful. If those that are successful occur above a certain threshold, they may become a national security issue. Thus, given the capabilities afforded by Information Age technologies and the increased dependence of the United States and other societies on them, one highly capable person pursuing a personal agenda could alter data, disrupt operations, or otherwise compromise critical information and communication systems.

Similarly, other established international actors, especially multinational corporations (MNCs) and nongovernmental organizations (NGOs), could pursue objectives that challenge or threaten a state's security. This is not a new phenomenon. But with capabilities afforded by emerging technologies, it is more possible for MNCs and NGOs to challenge and threaten a state's national security.

Third, virtual sources of challenges and threats to security will emerge as virtual international actors emerge. This will require innovative responses. Since emerging technologies will aid individuals and organizations with similar interests in widely scattered locations, a proliferation of the formation of "virtual" entities that stake claim to a role in domestic policy or international affairs will occur.

While most virtual sources of challenges and threats may be safely ignored by established international actors, some virtual entities who have strongly held views on issues and who have highly developed technological skills may become dangerous players on the national and international scene. It is not difficult to envision a technologically highly-capable radical splinter group of an ethnic, religious, or environmental movement acquiring a virtual identity and demanding that a state or corporation undertake a certain action or suffer extremely adverse consequences generated electronically from an unidentified remote site. Depending on the technical capabilities and credibility of the hypothetical virtual radical splinter group, such threats and demands could quickly become national security issues.

National security analysts and decision makers clearly have their work cut out for them if they are to successfully defend against cyber challenges and threats. They must differentiate between lower order challenges and threats and higher order challenges and threats to identify where the threshold is between national security and less pressing concerns. When a challenge or threat exists, they must identify from where it emanates, what its intent is, the degree and type of danger that it poses, and how to respond. They must do this in a strategic environment that is more complex than ever before.

New Security Frontiers: From the Three D's to the Whole of Government Approach and Beyond

According to many, including high officials in the past-Bush and present-Obama administration, the often inter-related and predominantly non-traditional nature of many of the challenges and threats to national and global security discussed in the preceding chapters and this conclusion require new approaches to national and global security. Traditional military capabilities will remain requisite to counter traditional military challenges and satisfy traditional military needs, but new thinking about security is needed if the twenty-first century world is to become safer and more secure.

During George Bush's administration (2001–2009), the concept of the "Three D's" was sometimes discussed, although admittedly in a vague and not clearly defined form, as the requisite way in which both traditional and non-traditional challenges and threats to security could be countered. Under the "Three D's," the United States would unite defense, diplomacy, and development to achieve a more peaceful and secure world. Thus, addressing the World Bank on July 17, 2001, President Bush said that the United States "must be guided by three great goals." The first was to keep the peace with military forces to support freedom and free states. The second was "to ignite a new era of global economic growth through a world trading system that is dramatically more open and more free." The third was "to work in true partnership with developing countries to remove the huge obstacles to development, to help them fight illiteracy, disease, unsustainable debt."[26]

Even after the September 11, 2001 attacks on the World Trade Center, and contrary to much popular belief, Bush continued to emphasize a non-explicit form of the "Three D's." For example, in his March 22, 2002 speech to the United Nations' "Financing for Development" Conference in Monterrey, Mexico, Bush declared his intention to increase US development assistance by 50 percent and create a "Millennium Challenge Account" to help developing states.[27] Similarly, President Bush in his introductory remarks to the September 2002 *National Security Strategy of the United States of America* document also referenced in oblique terms what later, in modified form, became known as the "Three D's":

> We will actively work to bring the hope of democracy, development, free markets, and free trade to every corner of the world. The events of September 11, 2001, taught us that weak states, like Afghanistan, can pose as great a danger to our national interests as strong states …
>
> The United States will deliver greater development assistance through the New encourage economic freedom. We will also continue to lead the world in efforts to reduce the terrible toll of HIV/AIDS and other infectious diseases …
>
> We are also guided by the conviction that no nation can build a safer, better world alone. Alliances and multilateral institutions can multiply the strength of freedom-loving nations. The United States is committed to lasting institutions like the United Nations, the World Trade Organization, the Organization of American States, and NATO as well as other long-standing alliances.[28]

Admittedly, even though the 2002 *National Security Strategy* document became more renowned as one of the basis for the so-called Bush Doctrine of Pre-Emptive Defense, and even though specific references to the "Three D's" can not be found in the document, it is clear that such an intent existed. The actual phrase "Three D's" was likely created in the later years of the Bush administration both for its alliterative qualities and its conceptual connections.[29]

Conversely, the "Three D's" concept has been criticized for its perceived failure to directly connect defense, diplomacy, and development. Critics of the "Three D's" asserted that boundaries remained between the three, and that these boundaries prevented effective implementation of policies designed to counter both traditional and non-traditional challenges and threats to security. Other critics avowed that defense, diplomacy, and development by themselves were not sufficient to cope with twenty-first century security challenges and threats, and that a more inclusive concept was needed.[30]

These criticisms were undoubtedly key factors that influenced the Obama administration to explicitly adopt a new approach to traditional and non-traditional security challenges and threats in its May 2010 *National Security Strategy* document, which declared that "a broad conception of what constitutes our national security" is needed,[31] and that the international order that the US sought to create could only be reached by:

> … resolv(ing) the challenges of our times- countering violent extremism and insurgency; stopping the spread of nuclear weapons and securing nuclear

> materials; *combating a changing climate and sustaining global growth; helping countries feed themselves and care for their sick; resolving and preventing conflict, while also healing its wounds* (emphasis added).[32]

But how to do this? Again according to the May 2010 *National Security Strategy* document, the answer is to adopt a "whole-of-government" approach to national and global security that views national and global security, as well as the challenges and threats to national and global security, in a comprehensive manner.[33] Thus, the May 2010 *National Security Strategy* document argues that a more holistic approach to national and global security must be developed and implemented.

Based in many respects on work previously done by the Project on National Security Reform, which had among its members Obama's National Security Adviser James L. Jones and Director of National Intelligence Dennis C. Blair, the whole-of-government approach, as recommended in the *National Security Strategy* document, requires development of a national-level strategic guidance document to address interagency roles and responsibilities and resolve issues between agencies at the seams of their operations. It also advocates an annual federal budget that contains both a whole-of-government approach to national security programs and the traditional department-by-department approach. What is more, the 2010 *National Security Strategy* document will require creation of a permanent framework that includes commonly understood strategic concepts and terms, widely-accepted operational principles, good working relationships between agencies, and a clear understanding of each agency's roles and responsibilities, especially where roles and responsibilities may overlap.

Even though it is better than what preceded it, the whole-of-government approach has at least two shortcomings.[34] First, it is not clear that a bureaucracy as massive as that of the US government can in fact operate on a whole-of-government basis. Bureaucratic silos have long been the enemy of more holistic approaches to issues that cross-bureaucratic jurisdictions, and while the 2010 National Security Strategy advocates overcoming the boundaries between silos, it does not lay out a roadmap as to how this may be accomplished. The bureaucratic rivalries and disagreements that separate even those arms of government within the Department of Defense and beyond that concentrate on national security and defense are legendary. It will be a tall task indeed to overcome them, even with a whole-of-government approach. It is doubtful that a simple directive can do the job.

Second, it is not clear that the whole-of-government approach is sufficient to achieve the security that is sought. Given the complexity and the inter-relationships of the challenges and threats to society that have emerged in the early twenty-first century – to reiterate, energy shortfalls, non-fuel mineral resources scarcity, food and fresh water shortages, encroachment of desertification, and cyber attacks – a "whole-of-society" approach broader than the "whole-of-government" approach that brings business and corporate interests, private individuals, non-government actors, and others into the effort, in addition to government, may be required to enhance twenty-first century security. This, however, will be an even more difficult task that creating a truly operational and successful whole-of-government approach.

To a certain extent, the 2010 *National Security Strategy* document recognizes both the importance and necessity of this broader beyond-the-government and whole-of-society approach. The document observers that "the American people and the private sector" must also be engaged in the promotion of security, and pledges the Obama administration to:

> support the development of prepared, vigilant, and engaged communities ... tap the ingenuity outside government through strategic partnerships with the private sector, nongovernmental organizations, foundations, and community-based organizations, [and] support [such partnerships] through enhanced opportunities for engagement, coordination, transparency, and information sharing.[34]

Even more specifically, the document recognizes that:

> new skills are needed to foster effective interaction to convene, connect, and mobilize not only other governments and international organizations, but also non-state actors such as corporations, foundations, non-governmental organizations, universities, think tanks, and faith-based organizations, all of whom increasingly have a distinct role to play on both diplomatic and development issues. To accomplish these goals our diplomatic personnel and missions must be expanded at home and abroad to support the increasingly transnational nature of twenty-first century security challenges. And we must provide the appropriate authorities and mechanisms to implement and coordinate assistance programs and grow the civilian expeditionary capacity required to assist governments on a diverse array of issues.[35]

Despite such good intentions, the development, implementation, and maintenance of such effective partnerships are easier said than done. Given their diverse and varied primary interests, will "corporations, foundations, non-governmental 0rganizations, universities, think tanks, and faith-based organizations" willingly participate in helping to identify and meet security challenges and threats? Will sufficient funding be available to develop required new skills? If such skills can be developed, will they then be applied to achieve security objectives, or other purposes? Despite such questions, the good news is that the intention is there.

The United States, and the world, then, are on the horns of a dilemma. Achieving twenty-first century security in light of the diverse challenges and threats that exist is and will be a challenging task that will grow more difficult with time. Innovative thought, policy experimentation, and unprecedented cooperation will be required across all sectors of society and all national and international actors. While the 2010 *National Security Strategy* document is a step in the right direction, there unfortunately is no guarantee that such thought, experimentation, and cooperation will be forthcoming or succeed. Even so, if the new and more broadly defined security frontiers of the twenty-first century are to be successfully defended, and if the new challenges and threats are to be successfully met, such thought, experimentation, and cooperation are necessities.

Notes

1 See *The Merriam-Webster Online Dictionary* at: http://www.merriam-webster.com/.
2 See *World Bank Development Indicators*, March 15, 2011.
3 See Steven Woehrel, "Russian Energy Policy Toward Neighboring States," Congressional Research Service Report for Congress, RL34261, September 2, 2009.
4 Erica Downs, *The Brookings Foreign Policy Studies Energy Security Series: China* (December 2006)
5 For discussions of "resource war" fears during the Cold War, see James Hensen, "High Strategic Stakes in Southern Africa," *National Defense* (May–June 1982), pp. 42–46; and Africa Report, Vol. 27, No. 4 (July–August 1982), p. 40. For a more balanced view of Soviet policies toward southern Africa during this period regarding non-fuel mineral resources, see Daniel S. Papp, *Soviet Policies Toward the Developing World During the 1980s: The Dilemmas of Power and Presence* (Maxwell Air Force Base, AL: Air University Press, 1986), pp. 294–7.
6 See MBendi Information Service for southern African production of these and other non-fuel mineral resources as a percentage of the world's 2009 total production. Available at: http://www.mbendi.com/indy/ming/af/p0005.htm.
7 Klaus Schulz and Judy Back, "Assessing the World's Undiscovered Non-Fuel Mineral Resources: A Cooperative International Project," *US Geeological Survey Mineral News* (September 2002, Vol. 1).
8 World Foresight Forum, *Rare Earth Elements and Strategic Mineral Priorities* (The Hague Center for Strategic Studies, 2010, p. 7). See also Jens-Dieter Becker-Platen, "Global Non-Fuel Mineral Resources and Sustainability, *US Geological Survey Circular 1294* (June 28, 2007), which discusses the availability of non-fuel mineral resources as renewable and non-renewable resources, the future availability of such resources, and the degree of resilience of the environment to the impact of mining.
9 For such a claim, see: http://www.redicecreations.com/article.php?id=11359. For one of many counter views, see Lisa Reisman, "Vast Riches of Minerals Found in Afghanistan Ought to Fall on Deaf Ears," *MetalMiner* (June 14, 2010). Available at: http://agmetalminer.com/2010/06/14.
10 See *Rare Earth Elements and Strategic Mineral Priorities*, p. 7
11 See Peter H. Gleick, *Water Conflict Chronology* (The Pacific Institute for Studies in Development, Environment, and Security, November 2008).
12 R. Clark and J. King, *Atlas da Agua* [*The Atlas of Water*] (Sao Paulo: Publifolha, 2005), p. 20.
13 Mike Hightower and Suzanne Pierce, "The Energy Challenge," *Nature* (March 20, 2008), pp. 285–6.

14 Ibid.

15 See, for example: BBC reports, June 2, 2000, at: http://news.bbc.co.uk/2/hi/middle_east; OECD Issues Brief, Development Assistance Committee, *Water and Violent Conflict*, (OECD: 2005); and Jonathan Watts, "Can the Sea Solve China's Water Crisis?," *The Guardian*, January 24, 2011. See also *Global Policy Forum* for a list of publications discussing past, present, and impending global fresh water shortages and crises. Available at: at http://www.globalpolicy.org/the-dark-side-of-natural-resources/water-in-conflict.html.

16 See, for example, Tim Gaynor and Steve Gorman, "Fast-Growing Western US Cities Face Water Crisis," *Reuters*, March 10, 2009). Available at: http://www.rivcoh2o.com/Portals/0/communityOutreach/newsArticles; and Andrew Freedman, "Report Warns of Worsening Western Water Crisis," *Climate Central: Sound Science and Vibrant Media* (February 14, 2011). Available at: http://www.climatecentral.org/blog/report-warns-of-worsening-western-water-crisis/.

17 See "Tri-State Water Wars History," Upper Chattahoochee Riverkeeper. Available at: http://www.chattahoochee.org/tri-state-water-wars-history.php.

18 See the United Nations Convention to Combat Desertification web-site. Available at: http://www.unccd.int/convention/text/convention.php?annexNo=0.

19 See *Desertification as a Source of Conflict in Darfur* (Washington, DC: Worldwatch Institute, 2011).

20 See Thomas Malthus, *An Essay on the Principle of Population* (New York: Oxford University Press, 2008).

21 See, for example, David McKeeby, "Many Modern Conflicts are Food Wars, Say Experts," America.gov, September 12, 2007. Available at: http://www.america.gov/st/develop-english/2007/September; and "IMF Head: Food Shortages Can Spark War," *CBS News*, April 13, 2008. Available at: http://www.cbsnews.com/video/watch/?id=4011418n&tag=related.

22 M.E. Kabay, "Attacks on Power Systems: Hackers, Malware," *Network World*, September 13, 2010, pp. 1–2.

23 *The Wall Street Journal*, April 8, 2009.

24 Glenn Derene, "How Vulnerable is US Infrastructure to a Major Cyber Attack?," *Popular Mechanics*, October 1, 2009. Available at: http://www.popularmechanics.com/technology/military/4307521.

25 For deeper discussions of these three implications, see Daniel S. Papp and David S. Alberts, "National Security in the Information Age," in David S. Alberts and Daniel S. Papp (eds), *Information Age Anthology: Volume II* (Washington, DC: C4ISR Cooperative Research Program, 2000), pp. 1–53.

26 George W. Bush, "Speech to World Bank," Washington, DC, July 17, 2001.

27 George W. Bush, "Speech to United Nations 'Financing for Development' Conference," Monterey, Mexico, March 22, 2002.

28 George W. Bush, "Introductory Remarks," in *National Security Strategy of the United States of America* (Washington, DC: US Government Printing Office, September, 2002).

29 Philip Zelikow, "Where Did the 'Three D's' Come From, Really," *Shadow Government: Notes from the Loyal Opposition*, February 29, 2009. Available at: http://shadow.foreignpolicy.com/posts/2009/02/23/.

30 See, for example, "The Three D's: Defense, Diplomacy, and Development," Center for American Progress, August 1, 2008. Available at: http://www.americanprogress.org/issues/2008/08/three_ds.html.

31 *2010 National Security Strategy* (Washington, DC: US Government Printing Office, May, 2010), p. 51.

32 Barack Obama, "Introductory Remarks," in Ibid.

33 Ibid., pp. 14–16.

34 *2010 National Security Strategy*, p. 16.

35 Ibid., pp. 14–15.

Index

9/11 20, 208

accessibility 4, 33, 34, 58–9
adaptability 85, 86
additionality 101–2, 103, 104–5
advanced agricultural era 35
Afghanistan 18, 20, 23, 122, 202
Africa 20, 35, 101, 125, 141, 181–2, 201–2, 203, 204; *see also individual countries*
agriculture 3, 6, 34–5, 72–5, 205; *see also* food production
air pollution 78, 97, 99
aircraft 115–16, 127
Algeria 16, 20
Allan, J.A. 152
Al-Mutaafie, Abdul Halim Ismail 178
aluminium 127
amity 149–50, 152, 165
Amoco-Arco 45
Anglo-Persian Oil Company 13, 37
Angola 20, 182, 185, 201
appliances 81, 104–5, 119
Arabian American Oil Company (ARAMCO) 15–16
Arafura Resources 121, 122, 123
Argentina 7, 20, 53, 128, 129, 151, 153–63
Armitage, Richard 23
Arrhenius, Svante 49
Asia 19, 20, 25, 36, 99, 141–2; *see also* individual countries
Atacama Desert 204
Australia 35, 99, 121–2, 123, 125, 128, 178
Avalon Rare Metals 121, 123
Azerbaijan 14, 22, 23, 25, 36

Bachler, G. 151
bagasse 72, 74–5
Baker, James 23
Baku-Novorossiisk pipeline 25
Baku-Supsa pipeline 25
Baku-Tblisi-Ceyhan pipeline 25, 26
Bangladesh 5, 68, 75–8, 82–3, 84–6, 205
banking 206–7
Barboza, Mario Gibson 159
baseline measurements 102, 103
batteries 117–18, 125, 127
Belarus 23
Belgium 129
Bennet, Andrew 87
Bielecki, J. 67
bilateral agreements 6, 14, 173–89
Bilgin, Mert 4
bio-diesel 53
bio-fuels 4, 6, 34, 40, 48, 53–5, 58, 72–5, 96, 114–16, 119, 129–30, 179
biogas 5, 68, 75–6, 77, 82, 83
biomass 70, 71, 78, 79
BKK 178
Blair, Dennis C. 210
Bolivia 15, 20, 122, 151, 153, 162–3
border disputes 154–5, 158
Boyd Orr, John 187
BP 45, 22, 34, 48, 51, 56, 58; *see also* British Petroleum
Brandt, Willy 22
Brazil
  carbon emissions 101
  economic growth 45
  ethanol 5, 53, 68, 72–5, 82–6, 114–15
  hydropolitics 7, 151, 153–63
  oil 20, 46, 48, 56
  rare earth elements 2, 121, 123
Brezhnev, Leonid 22
BRIC countries 45, 50–51; *see also* Brazil; China; India; Russia
Britain *see* United Kingdom
British East India Company 174

British Oil Administration 14
British Petroleum 16, 38; *see also* Anglo-Persian Oil Company; BP
Brzezinski, Zbigniew 18, 23
building codes 97, 104–5
Burundi 205
Bush, George H.W. 18
Bush, George W. 18, 20, 21, 199, 208–9
Buzan, Barry 145, 148–9, 150, 165

California 2, 121, 122, 123
Cambodia 178, 179
Canada 18, 99, 114, 121, 123, 125, 128, 129
carbon capture and storage (CCS) 96
carbon emissions 34, 41, 49–51, 53, 56–8, 69, 75, 82, 94–5, 96–7, 98–9, 100–102, 104–5
Carter, Jimmy 18
cars 11, 37, 43, 73–4, 82–3, 104–5, 117–19, 121, 125, 127
Caspian Pipeline Consortium pipeline 25
catalytic converters 120
cerium 120, 122, 124, 127
Certified Emissions Reduction credits 100
Chavez, Hugo 48
Chemetall 128
Cheney, Dick 20, 23, 24
Chevron 38, 45, 122; *see also* Standard Oil Company; Texaco
Chile 127–8, 129, 130, 151
China
  carbon emissions 50, 98, 101
  coal 35, 58, 201
  cook stoves 5, 68, 78–81, 82–3, 84–6
  desertification 204
  energy consumption 94, 98–9, 200, 201
  energy efficiency 98–9
  land deals 176–8, 181–2, 185, 187, 188–9
  mineral resources 2–3, 117, 118, 120–26, 127–9, 130, 203
  oil and gas 19, 20, 24–6, 45, 46, 48, 49, 56, 200, 201
China National Overseas Oil Corporation 122
China-Russia oil pipeline 24
Chinese National Petroleum Corporation 46, 48, 56
chromite 202
Churchill, Winston 37
Civil Society Agriculture Network 179
Clarke, R. 143
Clean Development Fund 105–6
Clean Development Mechanism (CDM) 93, 100–106
climate change 3, 4, 10, 19–20, 21, 34, 41, 49–51, 58, 93–107
climate engineering 96
climate policy 93–107
Clinton, Bill 18
coal 2, 10, 13, 32, 33, 34–5, 36, 42, 43–4, 58, 69, 79, 95, 201
cobalt 125, 126–9, 130, 202
Code of Conduct (food security) 173–4, 183–6, 187, 188–9
cogeneration 69, 71, 81–2, 83–4
cold fusion 42
Cold War 17–18
collective action theory 5
Colombia 151
colonialism 4, 9, 10, 12–14, 174, 179–80
compact fluorescent lamps (CFLs) 97, 120
conflicts of interest 182 ,185
Congo, Democratic Republic of 127–8, 129, 130, 176, 178, 205
Constantinople Agreement 14
construction industry 97, 104–5
Constructivist Theory 146–7, 165
Consultative Group on International Agricultural Research (CGIAR) 185
Cook, E. 34–5
cook stoves 5, 68, 75–6, 77–81, 82–3, 84–6, 97, 99
Copenhagen, School of 145, 147, 148, 150, 151, 152
Copenhagen negotiations 50, 98, 184
copper 122, 125, 128, 202
Corpus-Itaipu Tripartite Agreement 159–61
Costello, Peter 178
Council for Mutual Economic Assistance (COMECON) 22
coups d'etat 21, 181
critical raw materials 5–6, 113, 116–31
Cuba 22, 128
cyber security 7, 199, 205–8

Daewoo 181
defense 199, 208–9; *see also* military; warfare
deforestation 74–5, 78–9, 81, 96, 204
Denmark 5, 68, 69–72, 81–2, 83–6
deprivation 180–81; *see also* poverty
desalination 52
desecuritization 146, 149
desertification 7, 199, 203, 204–5
developing countries 2–3, 5, 6, 21, 50–51, 54, 93, 97, 98–9, 100, 101, 102–3, 104, 173–89, 205; *see also individual countries*
development 199, 208–9
diamonds 202
diplomacy 199, 208–9
diversification
  of energy sources 5, 6, 33–4, 41–5, 56, 68, 94, 113, 114–15
  of suppliers 6, 17, 25, 26, 113, 114, 129
Drake, Edwin 36
drought 3, 151; *see also* water scarcity
Du Plessis, A. 144
dysprosium 118, 121, 123, 124

East China Exploration 122
Eastern Siberia-Pacific Ocean pipeline 24–5
economic crisis 20, 44
economic growth 44–5, 56, 113
economic security 5–6, 10–11
Ecuador 20, 151
Egypt 205
electric cars 118–19, 121, 127
electricity consumption 34, 97, 102
electricity generation 36, 44, 69–72, 74–5, 95
electricity supply 5, 68, 69–72, 81–2, 83–6
Electrobras 155
Elhance, A.P. 142
embargos
  oil 16, 38–40, 69, 72, 125
  rare earth elements 124–5, 203
Emergent Asset Management 178–9
end-use efficiency 102–3, 104–5
energy consumption 2, 34–5, 42–4, 56, 70, 93, 94, 97, 98–9, 102, 200–201
energy efficiency 69–70, 77–8, 79, 81, 93, 96–100, 101–6, 120
energy governance 3, 5, 21–2, 99–100
energy independence 6, 17, 94, 99; *see also* self-sufficiency
energy intensity 98–9
energy markets 31–2; *see also* oil markets
energy mix 33–4, 41–5, 94
energy politics 13–26, 36–40; *see also* geopolitics
energy prices 31, 95; *see also* oil prices
energy security 1–2, 3–5, 9–13, 16–22, 26–7, 31–3, 38, 40–59, 67–86, 93–107
energy shortfalls 200–201
energy transitions 4, 31–59
enmity 149–50, 151, 152, 165
environmental issues 20–21, 35, 41, 42–3, 49–51, 53, 54, 55, 56–8, 73, 203; *see also* carbon emissions; climate change; deforestation; desertification; oil spills
equity 81–2, 86
erbium 123
ethanol 5, 53, 54, 68, 72–5, 82–3, 84–6
Ethiopia 176
Europe 23, 24, 25, 26, 35, 53, 71, 114, 121, 124, 125, 130, 143, 188, 201; *see also individual countries*
European Commission 125
European Union 23, 24, 25, 53, 71, 121, 124, 125, 130, 188
europium 120, 123
export bans 177, 188; *see also* embargos
export quotas 123–4, 125
Extractive Industries Transparency Initiative (EITI) 187–8
Exxon 38; *see also* Exxon Mobil; Standard Oil Company
Exxon Mobil 45, 56
Exxon Valdez 51, 56

Fairgrieve, James 10, 13
famine *see* food shortages
FAST principles 4, 33, 34, 58–9
feasibility 4, 33, 34, 58–9
Ferraz, Otávio Morcondes 155
Fiat 74
Figueiredo, João Batista 159–60
financial crisis 20, 44

financial markets 20
Finland 71, 128
first oil regime 13–16
First World War 10, 11, 37
Florida 123
fluid cracking catalysts 120
FMC 127
Foley, Meraiah 5, 93–107
Food First Information Network 179
food prices 6, 55, 173, 175, 181, 205
food production 3, 6, 173–89, 205
food security 1, 3, 6, 34, 40, 53–5, 58, 96, 173–89
food shortages 199, 205
Ford 37, 74, 118
Ford, Henry 37
foreign dependency 5–6
foreign direct investment (FDI) 25, 173, 174–89
fossil fuels 2, 4, 33–40, 41, 42–3, 49, 55, 95, 99; *see also* coal; gas; oil
France 14–15, 53, 99, 114, 129
Franco-Prussian War 201
fuel cells 42, 52
fuel shortages 79
fuel-wood 78–9, 81
Fukuyama, Francis 19

gadolinium 123
gallium 126–7, 129
gas 20, 21, 22–6, 27, 42, 43, 46, 48, 52, 55, 69–70, 71, 95
gas-for-pipes deal 22
gasohol 73
Gazprom 23, 27, 46, 48, 56
Geisel, Ernesto 156
General Motors 118
genetically modified (GM) crops 182
geopolitics 3–4, 9, 11–13, 17–22, 26–7, 36–40, 46–9, 154–6
George, Alexander 87
Germany 11, 12–13, 14, 22, 37, 53, 71, 114, 120, 124, 129, 130, 201
global warming *see* climate change
globalization 3, 4, 19–22
Gobi Desert 204
gold 202
Goldthau, A. 23
Goulart, João 155
government funding 119, 126; *see also* subsidies
government legitimacy 181
Grain 179
Grameen Shakti 75–8, 82–3, 84–6
Great Game 23–4
Great Western Minerals 121
greenhouse gas emissions *see* carbon emissions
Greenland 121
Guerreiro, Ramiro Saraiva 159–60
Gulbenkian, Calouste Sarkis 36, 37
Gulf Oil 16, 38
Guyana 151

Haiti 205
Hankey, Maurice 14
Harcourt, Viscount 14
Haushofer, Karl 12, 13, 23
Hawaii 35
health issues 35, 78, 81
Hensel, Nayantara 5–6, 113–31
Heterogeneous Complexes 149, 165
Hinrichsen, D. 143
Hitler, Adolf 13
holmium 123
Homogeneous Complexes 149
Honda 118
Hong Kong 24
household energy use 5, 68, 75–8
Hoyos, C. 46
hunting era 34
hybrid cars 43, 117–19, 121, 125, 127
hydro-energy 41, 42, 71, 74, 154–61
hydrofluorocarbons (HFCs) 101
hydrogen 41, 42, 52, 55
hydrological interdependence 144, 145, 152
Hydropolitical Complexes (HCs) 145, 152, 165
hydropolitical interdependence 142
Hydropolitical Security Complexes (HSCs) 7, 151–2, 165
hydropolitics 7, 139–45, 151–65

Idaho 123, 128
Iguaçu Act 156
Illia, Arturo 155

imperialism *see* colonialism
incentives 83–4, 86, 104–5; *see also* subsidies
inclusivity 82–3, 86
India 2, 26, 35, 45, 50, 94, 98, 99, 101, 121, 125, 177, 200, 201, 203
indium 126–8
Indonesia 13, 45, 53
industrial era 35
industrial revolution 35
industrialization 10, 35
institutions 31–2
Intergovernmental Panel on Climate Change (IPCC) 96
internal combustion engines 36; *see also* cars
International Atomic Energy Agency (IAEA) 52
International Energy Agency (IEA) 18–19, 96–7, 98, 102, 106
International Energy Program (IEP) 19
International Food Policy Research Institute (IFPRI) 173, 176, 183, 185
Iran 2, 13, 14, 18–20, 27, 40, 46, 48, 49, 52–3, 56, 58, 95, 180
Iran-Iraq War 18, 19, 40
Iranian Revolution 18, 40
Iraq 2, 13, 14, 16, 18, 19, 20, 40, 48, 114, 151, 152, 200
Iraq Petroleum Company 37; *see also* Turkish Petroleum Company
iron 122, 201
Israel 40, 53
Itaipu Binational Power Plant 154, 158–9
Italy 15

Japan
  carbon emissions 50
  coal 35
  energy consumption 201
  energy efficiency 99
  mineral resources 120, 121, 124, 125–6, 128, 129, 130, 202, 203
  oil and gas 11, 24–5, 26, 37
  Second World War 37–8
Jarch Capital 178
Jones, James L. 210

Kazakhstan 22, 23, 25–6
Kazakhstan-China oil pipeline 25
Kenya 176, 179
Kenya Nature 179
Keohane, R.O. 142, 156
King, J. 143
Kissinger, Henry 23
Kjellen, Rudolph 12
Kosygin, Alexei 22
Krasner, S. 17
Kuwait 18, 40, 114, 176, 180, 200
Kyoto Protocol 19, 49–50, 98, 100, 101, 103, 106–7, 184

land deals 6, 173–89
Language Theory 147
lanthanum 120, 122, 125, 127
Las Leñas Agreement 163
lasers 119–20
Latin America 13, 20, 35, 114, 125, 128–9, 174; *see also individual countries*
Le Prestre, P. 143
lead 128, 202
Lebanon 14
Lenin, Vladimir Ilyich 10
Libya 16, 27, 52
Licadho 179
light-emitting diodes (LEDs) 120, 127
lighting 97, 102, 104, 105, 117–18, 120
linear fluorescent lamps 120
lithium 122, 125, 126–9, 130–31
Lopez, Solano 154
lutetium 123
Lynas Corporation 121, 122, 123, 125

Mackinder, Halford 12–13, 23
Madagascar 181
Magalhães, Juracy 155, 156
magnets 117–18, 119–20, 125–6
Malaysia 46, 48–9, 53, 56, 121, 123
Malthus, Thomas 205
manganese 125, 127
market deregulation 19, 20
mass production 11
Mercosul 161, 163
Mexico 15, 18, 40, 101
Michigan 128

microcredit 75, 76, 84
Middle East 13–14, 16, 18, 37, 38, 95, 98–9, 114, 141, 151, 200, 203; *see also individual countries*
military
  use of rare earth elements 119–20, 126
  use of oil 10, 11, 36–7, 115–16
  see also defense; warfare
mineral resources 2–3, 6, 116–31, 201–3
Minnesota 128
Mobil 38; *see also* Exxon Mobil; Standard Oil Company
Molchanov, Mikhail 3–4, 9–27
Molycorp 121, 122, 123, 126
Mongolia 123, 125, 204
monitoring 83, 86, 104–5
Montana 123
Mozambique 182, 201
Murphy, S. 182

Nabucco gas pipeline 25, 26
Namib Desert 204
Namibia 201
National Iranian Oil Company (NIOC) 46, 48, 56
National Security Strategy (US) 209–12
nationalization 15, 16, 20
natural gas *see* gas
navies 10, 11, 36–7, 115–16
neo-colonialism 10–11, 179–80, 182, 185
neodymium 118, 119–20, 122, 127
neoliberalism 19–20
Netherlands 13, 71
Nevada 128
New Energy Order 33–4, 40–55, 56–7
new seven sisters 45–9, 56
New Zealand 35
nickel 125, 127, 128, 202
Niemetz, Hans 11
Nigeria 2, 40, 180
Nkumrah, Kwame 179–80
Nobel, Ludwig 36
Nobel, Robert 36
Nogues, Alberto 160
North Carolina 128
North Korea 52
Northern Uranium 122
Norway 71, 129
nuclear energy 4, 33, 34, 40, 43, 48, 51–3, 55, 56, 58, 96, 206
Nuclear Non Proliferation Treaty 52
nuclear weapons 34, 51, 52, 56, 58
Nye, J. 142

Obama, Barack 18, 199, 209–11
Ocheje, P. 188
off-grid energy services 5, 68, 76
oil
  dependence upon 43, 95
  and energy politics (Russia) 22–6
  energy transition to 32, 33, 35, 36–40, 42
  first oil regime 13–16
  second oil regime 16–19
oil companies 21–2, 23, 36–7, 45–9; *see also individual company names*
oil crises 4, 9, 17–18, 38–40, 69, 72, 200
oil drilling 36
oil embargos 16, 38–40, 69, 72, 125
oil exploration 51, 56, 58
oil markets 5, 16–17, 19
oil pipelines 23–6, 27
oil prices 2, 5, 16, 22, 25, 40, 43, 45, 51, 53, 54, 55, 72, 200
oil production 2, 9, 13–22, 45–9, 200
oil rents 21
oil reserves, strategic 17
oil resources 1–2, 13–22, 180
oil spills 22, 34, 48, 51, 56, 58
O'Neill, Jim 44–5
Organization for Economic Co-operation and Development (OECD) 18, 44–5, 56, 98
Organization of Petroleum Exporting Countries (OPEC) 16, 17, 18, 38, 40, 69, 72, 114, 200
organizational multiplicity 84–5, 86
overlay 149
Overseas Development Aid (ODA) 178

Pakistan 52, 122
Pande, Kabinga 178
Papp, Daniel S. 7, 199–212
Papua New Guinea 179
Paraguay 7, 151, 153–63
Paraguay-Parana waterway 154, 161–4

Paraná River 153, 154–5, 159
Partners in Melanesia 179
passive energy housing 97
Pastor, Carlos Washington 160
Pastor, Sapena 156
Pearl Harbour 37, 125
Perle, Richard 23
Peru 128, 151
Petrobras 46, 48, 56
Petróleos de Venezuela 46, 48, 56
petroleum *see* oil
Petroleum Administration for War (PAW) 15–16
Petroleum Reserves Corporation 15–16
Petronas 46, 48–9, 56
Philippines 35, 178
Pinto, T.G. 158
pipelines 23–6, 27
platinum 202
political violence 21, 181
politicization 145–6
population growth 2, 3, 131, 177, 189, 203, 205
potassium 202
poverty 1, 55, 75, 78; *see also* deprivation
praseodymium 118, 123, 127
Prata Basin 7, 145, 153–65
Prata Basin Treaty 157–8
primitive agricultural era 34
primitive era 34
Principles for Responsible Agricultural Investment 183–4
privatization 22–3
Proálcool program 72–5, 82–3, 84–6
programmatic CDM 104, 105
Putin, Vladimir 26

Qatar 176
quantitative easing 124
Queiroz, Fábio Albergaria de 6–7, 139–65

railways 36
Rajeolina, Andry 181
rare earth elements 2–3, 6, 116–26, 128–31, 202–3
Ratzel, Friedrich 10–11
Ravalomanana, Marc 181
Reagan, Ronald 18
Red Line Agreement 37
Regional Security Complexes (RSCs) 145, 148–52
renewable energy 5, 33, 41, 42, 55, 68, 69–72, 75–7, 95, 114–15, 119; *see also* bio-fuels; hydro-energy; solar energy; tidal energy; wind energy
resource competition 1–4, 19, 21
resource depletion 4, 10, 19
resource nationalism 19–20
Rockefeller, John D. 36
Roosevelt, Franklin D. 15
Rothschild family 36
Royal Dutch Shell 13, 15, 16, 36, 37, 38, 45, 48, 56
Ruggie, John 22
Russia
  carbon emissions 50
  mineral resources 121, 123, 125, 129, 202
  oil and gas 2, 4, 9, 10, 13, 14, 15, 20, 22–6, 36, 40, 45, 46, 48, 49, 56, 95, 114, 180, 201
  see also Soviet Union

Sahara Desert 204
Sahel Desert 204
samarium 123, 126
Samuel, Marcus 36–7
San Remo Agreement 15
Sanguinetti, J. 163
Sassou-Nguesso, Denis 178
Saudi Arabia
  land deals 176–8, 180, 181, 185, 187, 188–9
  oil 13, 15–16, 18, 26, 27, 38, 46–8, 56, 114
Saudi Aramco 46–8, 56
Schulz, M. 151–2
Schumacher, E.F. 67
Scowcroft, Brent 23
second oil regime 16–19
Second World War 13, 15–16, 37–8, 125, 201
sectoral CDM 104–5
securitization 6, 145–8

Security Complex Theory 7, 148–51
self-sufficiency 12, 13, 17; *see also* energy independence
Semple, Ellen Churchill 10–11
Semyenov, F.N. 36
Serageldin, Ismail 143
seven sisters 16–17, 34, 38, 45–6; *see also* new seven sisters
Shalabi, Tarek 178
Shanghai Cooperation Organization 24
Sheenan, M. 147
Shell *see* Royal Dutch Shell
Shepherd, Ben 6, 173–89
shipping 10, 11, 35, 36–7, 115–16
Silveira, Azerado de 156
Six Day War 16
Smolansky, O.M. 23
Sojitz Corporation 125
solar energy 5, 41, 42, 68, 75–7, 81, 82–3, 84, 119
Somalia 205
South Africa 2, 45, 121, 201
South American Regional Security Complex 7, 145, 150–51, 153–65
South Korea 25, 26, 99, 101, 125, 128, 176, 181, 18, 188
Sovacool, Benjamin 4–5, 67–86
Sovereign Wealth Fund (SWF) 177
sovereignty 181–2
Soviet Union 15, 18, 19, 22, 40, 201–2; *see also* Russia
speech-acts 146–8
Speildoch, A. 182
SQM 128
Standard Oil Company 13, 16; *see also* Chevron; Exxon; Mobil
Statoil 56
steam power 35, 36
Stern, J.P. 23
stockdraw 18–19
strategic oil reserves 17
strategic resources 4, 10
Stroessner, Alfredo 155, 156
subsidies 22, 53, 69, 70–72, 74, 79–80, 84, 85, 104–5; *see also* government funding; incentives
Sudan 48, 176, 178, 204, 205
Suez crisis 16
sugarcane 72–5
Sungwaya Welfare Association 179
super alloys 127
sustainability 4, 33, 34, 49, 58–9, 100, 104–5, 183
Sweden 71
Sykes-Picot Agreement 14
Syria 14, 151, 152

Tailson Minerals 127–8
Taiwan 26
Tanzania 182
taxation 69–70, 73, 81–2, 124
technological era 35
tellerium 126–7, 128–9
terbium 118, 120, 121, 123
terrorism 20, 21, 199
Texaco 38, 45; *see also* Chevron
Texas 36
Thailand 53
Theory of Speech-Acts (TSA) 146–7
thorium 121, 123
three Ds approach 199, 208–9
thulium 123
tidal energy 42
Tigris-Euphrates basin 151
tin 128
titanium 202
Total 56
Toyota 118, 125
transparency 4, 33, 34, 58–9, 183, 187–8
transport
  aircraft 115–16, 127
  cars 11, 37, 43, 73–4, 82–3, 104–5, 117–19, 121, 125, 127
  and efficiency 97, 102–3, 104–5
  and fuel 35, 68, 72–5
  railways 36
  shipping 10, 11, 35, 36–7, 115–16
Treaty of Itaipu 156, 158–9
Treaty of St Petersburg 14
tungsten 122
Turkey 13, 14–15, 151, 152
Turkish Petroleum Company 14, 15, 37; *see also* Iraq Petroleum Company
Turkmenistan 23, 25, 48, 52

Turkmenistan-China gas pipeline 25
Turton, A.R. 152

Uganda 176
Ukraine 23, 24
United Fruit Company 174
United Kingdom 11, 13–16, 23, 24, 25, 35, 36–7, 114, 143, 178–9, 202
United Nations Conference of the Parties (COP) 98
United Nations Convention to Combat Desertification 204
United Nations Food and Agriculture Organization (FAO) 6, 173, 174–5, 176, 183, 184–5, 187, 189
United Nations Framework Convention on Climate Change (UNFCCC) 100
United States
- agriculture 188
- biofuels 114–16
- carbon emissions 50
- coal 35
- cyber security 206
- desertification 204
- energy independence 6, 99, 113–31
- ethanol 5, 53, 84
- land deals 178
- non-rare earth minerals 126–9, 202
- oil 2, 6, 11, 13, 14–21, 23, 24, 25, 26, 36, 40, 200–201
- rare earth minerals 116–26, 128–31
- security policies 208–12
- water resources 203

Unocal 122
uranium enrichment 52
Uruguay 151, 153–4, 163
USSR *see* Soviet Union

Venezuela 2, 18, 20, 27, 40, 46, 48, 49, 56, 114, 151, 180, 205
Vietnam 22, 123, 125
Volkswagen 74

Waever, Ole 145, 147, 148, 149, 150
Waltz, K. 156
warfare 10, 11, 17, 37–8, 143, 199, 201, 205
water distribution 6–7, 139–42
water resources 3, 6–7, 139–45, 151–65, 203–4
water scarcity 141, 142, 143, 199, 203–4, 205; *see also* drought
water stress levels 141
WeltHungerHilfe 179
Weltwirtschaft und Entwicklung 179
Wendt, A. 146
whole-of-government approach 199, 210–12
Wilde, J. 149
Willrich, M. 17
wind energy 41, 42, 69–72, 81–2, 83–4, 85, 117–18, 119
wood *see* fuel-wood
Woods, N. 147
World Bank 173, 183, 185
World Food Board 187
World Food Program 176
World Trade Organization 124, 188

Yap, Arthur 178
Yeltsin, Boris 24
ytterbium 123
yttrium 119, 120, 123

Zaire 201
Zambia 178, 182, 185
zero energy buildings 97
Zimbabwe 182, 205
zinc 128, 202

For Product Safety Concerns and Information please contact our EU representative GPSR@taylorandfrancis.com
Taylor & Francis Verlag GmbH, Kaufingerstraße 24, 80331 München, Germany

www.ingramcontent.com/pod-product-compliance
Ingram Content Group UK Ltd.
Pitfield, Milton Keynes, MK11 3LW, UK
UKHW011005180926
475701UK00004B/570

* 9 7 8 1 1 3 8 2 5 0 4 1 3 *